Safer City Cen

FLORIDA STATE
UNIVERSITY LIBRARIES

JUN 0 5 2001

TALLAHASSEE, FLORIDA

Safer City Centres
Reviving the public realm

edited by

Taner Oc and
Steven Tiesdell

HT
178
.G7
S24
1997

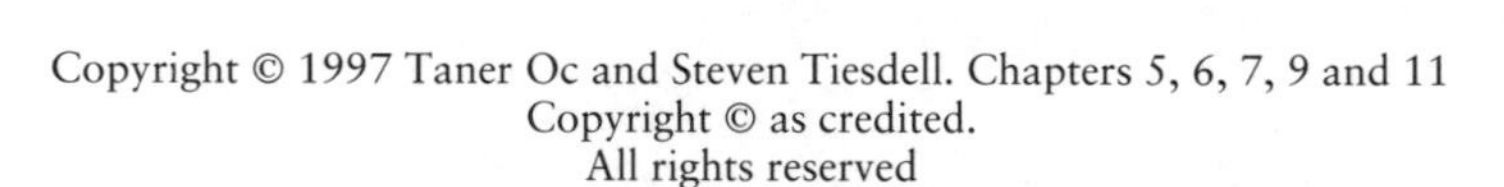

Copyright © 1997 Taner Oc and Steven Tiesdell. Chapters 5, 6, 7, 9 and 11
Copyright © as credited.
All rights reserved

Paul Chapman Publishing Ltd
144 Liverpool Road
London
N1 1LA

Apart from any fair dealing for the purposes of research or private study, or criticism or review, as permitted under the Copyright, Designs and Patents Act, 1988, this publication may be reproduced, stored or transmitted, in any form or by any means, only with the prior permission in writing of the publishers, or in the case of reprographic reproduction, in accordance with the terms of licences issued by the Copyright Licensing Agency. Inquiries concerning reproduction outside those terms should be sent to the publishers at the abovementioned address.

British Library Cataloguing in Publication Data

Safer city centres : reviving the public realm
1. Urban cores – Social aspects 2. City planning – Social aspects
I. Oc, Taner, 1944– II. Tiesdell, Steven
711.4

ISBN 1 85396 316 X

Typeset by Dorwyn Ltd, Rowlands Castle, Hants
Printed and bound in Great Britain

A B C D E F G H 9 8 7

Contents

Acknowledgements

This book is a culmination of work undertaken by the authors since 1990. We were invited by the Safer Cities Project in Nottingham to study women's perception of safety in the city centre. We organised interactive meetings bringing together planners, politicians, service providers and women's groups. To our surprise no bus operators attended the first meeting. From the start it was quite clear that if city centres were safer for women, they would be so for everybody. It was also clear that some policymakers and service providers did not appreciate this. We are pleased to note that since the initial meeting, significant changes of attitude have taken place.

We would like to acknowledge the contribution of our former colleague, Sylvia Trench, in the earlier stages of this work. Indeed, she has continued her interest in safer transport and has a chapter in this book. We are grateful to Nottingham Safer Cities Project, Nottingham City Council and Nottinghamshire County Council for their support of various dimensions of the work, especially the interactive workshops and the survey undertaken by Yasmine Guessoum-Benderbouz. We would also like to express our gratitude to Boots The Chemists and Chris Hollins, who continues to be a very supportive friend. Martin Garratt, Jane Ellis, Peter Collard, Mark Nicholls and Barry Cox gave us their valuable time – we thank them. Finally we would like to thank Linda Francis, Sarah Shaw and Jenny Chambers, who helped with typing and proofreading, Steven Thornton-Jones, who helped with illustrations and Glynn Halls, who helped with photographs.

Figure 2.1 is reproduced from Duncan, S. (1977) Mental maps of New York, *New York Magazine*, December 19, pp. 51–72.

Figure 1.2 is reproduced from Davis, M. (1992) Beyond Blade Runner: Urban Control – the Ecology of Fear, Westfield: Open Magazine Pamphlet Series.

Taner Oc
Steven Tiesdell

List of Contributors

CHRIS HOLLINS
Chris Hollins works within the Town Centre Management Department of Boots The Chemists and has been actively involved with the Town Centre Management movement since his role as Scotland's first Town Centre Manager in Falkirk in 1989. He is currently Chairman of the Association of Town Centre Management and provides guidance and insight into good practice in the concept for many of the country's principal cities.

TIM HEATH
Tim Heath is an architect-planner and a Lecturer in Planning in the Department of Urban Planning, University of Nottingham. He holds degrees from the University of Manchester and the University Nottingham and has research and teaching interests in urban regeneration, urban design and computer aided design. He co-authored *Revitalizing Historic Urban Quarters* (Oxford: Butterworth-Heinemann).

STEPHEN LISTER
Stephen Lister is a former senior police officer who specialised in training, communications, research, computer applications and organisational issues throughout a career with the Derbyshire police spanning more than 36 years. He has worked at local, national and international level. During his last six years of service he was seconded as Deputy Senior Police Advisor to the Home Office Police Research Group – a research unit staffed by researchers, scientists and civil servants. He co-authored each bi-annual publication of *Register of Policing Research* (London: HMSO). Since retiring from the police service he has become a self-employed consultant.

TANER OC
Taner Oc is the Head of the Department of Urban Planning at the University of Nottingham. He was educated at Middle East Technical University, the University of Chicago and the University of Pennsylvania and holds degrees in architecture, planning and sociology. He has teaching and research interests in urban design and urban regeneration. He co-edited *Current Issues in Planning Volume I* (Aldershot: Gower) *Current Issues in Planning Volume II*

(Aldershot: Avebury), and co-authored *Urban Design: Ornament and Decoration* (Oxford: Butterworth-Heinemann) and *Revitalizing Historic Urban Quarters* (Oxford: Butterworth-Heinemann). He is also the founding editor of the *Journal of Urban Design*.

ROBERT POOLE
Robert Poole is a Detective Chief Inspector in the West Midlands Police specialising in crime intelligence. He is an international authority on retail crime prevention and lectures extensively on the subject. He won the Ernst & Young Police Foundation Award in 1992 in relation to this work. He is the author of *Safer Shopping: Identifying Vulnerability in Shopping Malls* (Boston: Butterworth Heinemann).

ROBERT STICKLAND
Robert Stickland is currently Assistant City Centre Manager in Leicester. He holds a degree in geography from the University of Liverpool and a degree in planning from the University of Nottingham.

STEVEN TIESDELL
Steven Tiesdell is an architect-planner and a Lecturer in Planning in the Department of Urban Planning, University of Nottingham. He was educated at the University of Nottingham and has research and teaching interests in urban regeneration, urban design and housing. He co-authored *Urban Design: Ornament and Decoration* (Oxford: Butterworth-Heinemann) and *Revitalizing Historic Urban Quarters* (Oxford: Butterworth-Heinemann). He is also the Book Reviews and Practice Notes editor of the *Journal of Urban Design*.

TIM TOWNSHEND
Tim Townshend holds degrees from the University of Newcastle and the University of Manchester. He is a Lecturer in Urban Design and Conservation, at the Department of Town and Country Planning, University of Newcastle. He is a member of the Department's Environment and Safety Research Group who have produced a series of papers, including one by the author on Lighting, Crime and Safety.

SYLVIA TRENCH
Sylvia Trench was formerly a lecturer in the Institute of Planning Studies (now the Department of Urban Planning) at the University of Nottingham. She previously worked at the Political & Economic Planning Unit, the National Economic Development Office and the Department of Economic Affairs. She co-edited *Current Issues in Planning Volume I* (Aldershot: Gower) and *Current Issues in Planning Volume II* (Aldershot: Avebury). In recent years, her research has mainly concerned public transport and she has contributed articles to various journals. Since 1993 she has been working as a freelance lecturer and consultant on transport and planning.

Preface

On Saturday 17 July 1993 several hundred women took part in an evening demonstration organised by STRIDE (Safe Travel for Women) in Nottingham's Market Square. Their aim was to reclaim the streets at night from unruly, young and predominantly male revellers who make it unpleasant and unsafe. As reported in the *Nottingham Evening Post*, they vowed that women 'should continue to fight to reclaim the space which is rightfully theirs. Women spend all their lives restricting their movements . . . because of fears for their personal safety.'

The marchers called on the city and county councils to take immediate action to make the city's streets safer for women; safer streets for women in effect means safer streets for everyone. The demonstration echoed similar 'Reclaim the Night' marches that had occurred in Italy and elsewhere in Europe in the late 1970s, but in the intervening years our city streets and city centres have become less rather than more safe.

The women demonstrating in Nottingham were telling the authorities that they find the city centre an alien space: a place which they did not feel able to use and to which they did not go. As the city centre was no longer pleasant and safe, they were restricting their movements and their choices. The demonstration was an attempt to reclaim their legitimate rights as citizens to use the city centre without fear of unpleasant experiences and personal danger. The lack of women and other social groups is an impoverishment of the public realm, the life of the city, and a denial of the notion of cities as places of civilisation.

The above demonstration and a number of surveys (**see Table 2.4**) show that increasing numbers of women – and many middle aged men – consider city centres in Britain to be dangerous places, especially at night. At night the problem is either that city centres are deserted or that they are dominated by the 'wrong kind of people'. As early as the 1970s, surveys of the use of downtown areas of US cities (ISR, 1975) identified fear of crime as a critical variable in the diminished use of in-town commercial facilities. More generally there is a real concern that fear of crime will result in the loss of a 'public' life in cities. In the competition for the scarce spatial resources of the city, the young, unruly and criminal – and predominantly male – are wresting control of the city centre from the rest of the population. As a result those with choice are electing not to use the city centre so entraining a spiral of decline which makes the situation worse for all social groups. An editorial in the *Nottingham Evening Post* (1992, p.4) cogently outlined the issue:

> The researchers, the academics, the quango-types and the politicians can call it a 'perception' if they like. They can claim that 'the fear of violence in Nottingham city centre at night is far greater than the reality' as often as they like. Those are merely words. The truth is that thousands of people will not go into the city centre at night if they can avoid it, particularly at weekends. And the reason for that is that they feel intimidated by large groups of noisy, often drunken, youths. There might be odds of ninety-nine to one against being attacked. But they would rather not take the chance. And at best they are offended by the behaviour of yobs. It's not everybody's idea of entertainment to have to observe obscene chanting, swearing, belching, vomiting bunches of brain-dead lager louts lurching from pub to pub and stopping for the occasional punch-up en route. And despite official claims to the contrary the yobs do rule the city centre at night because most people have abdicated their right to a pleasant evening out for fear of running into them.

The diminishing use of the city centre has in some cities been paralleled by an increasing militarisation of both its architecture and its police force raising concerns about both the kind of city and society being created. The danger is that the continuation of these trends will mean 'fortress' cities whose citizens emerge heavily armed from defensive bunkers, fortified with barbed wire, heavy padlocks, bars, guard dogs and private security forces, to scuttle about in abject fear of their fellow citizens. Davis (1990, p.224) argues that Los Angeles is already a fortress city 'brutally divided between "fortified cells" of affluent society and "places of terror" where the police battle the criminalised poor.'

In Britain, as in the USA, a political emphasis on 'rugged individualism' and on the private sector rather than the state has encouraged individual self-help solutions and approaches. As Bottoms (1990, p.20) argues:

> If individualism really is unstoppable, the end result . . . could ultimately be a society with massive security hardware protecting individual homes, streets, and shops . . . But even if this heavy investment in defensive technology were to decrease crime rates, all the evidence suggests it would increase people's fear of crime.

As well as the fortress city, there is the 'panoptic city'. The idea originates in Michel Foucault's discussion of Bentham's Panopticon – an all-seeing architectural form designed to keep prisoners under constant surveillance. It is the extension of this concept beyond buildings, to public spaces, city centres and even whole cities, that ushers in the related spectre of 'Big Brother' forms of oppressive state control.

The overarching issue for this book is whether all cities in the twenty-first century have to be fortresses with urban life endured under sufferance, dictated and circumscribed by fear? Or whether there are more humane and positive alternatives whereby urban living and urban life can be enjoyed? As such it addresses the fundamental issue of how we as a society want to live in cities and what kind of public life exists in city centres.

This book discusses safety in the public space of city centres and, thus, the public realm. By implication it is also concerned with public order and disorder. Rather than with major social and/or politically motivated disruptions, such as riots, it is concerned with 'everyday' breaches of that order. Its concern is with certain types of crime, such as personal crime. It is particularly concerned with those crimes which affect people's perceptions of safety in – and their use – of city centres, and is therefore concerned with crime which is 'visible' and 'on the street'. As a consequence, it is not directly concerned with retail crime which has been covered elsewhere (see for example Poole, 1991; 1994; Beck & Willis, 1995).

The various chapters discuss how crime and especially fear of crime could be reduced and how the centres of towns and cities might be reclaimed, made safer and more liveable. Its emphasis is on cities rather than on the more abstract concerns of criminology or sociology, although these disciplines make important contributions. It is therefore firmly within the remit of urban planning and urban management, seeking to raise the profile of and emphasis on safety issues. To this end, it is edited – and largely written by – architects and planners with important contributions from town centre managers and the police. Its central concern is with the design and management of city centres and how this can be modified to reduce opportunities for anti-social behaviour or crime and lessen fear of crime.

The opening chapter discusses the general social and economic trends that are changing cities and city centres. Its title is an allusion to Jane Jacob's seminal book, *The Death and Life of Great American Cities*, which significantly bore the sub-title, *The Failure of Town Planning*. The second chapter discusses the different dimensions of the criminal act (legal, motivation/offender, victim and opportunity). Chapter 3 outlines the techniques of opportunity reduction approaches to crime prevention, while Chapter 4 represents the broader concept of a Safer Cities approach focusing on partnerships and an emphasis on the more vulnerable members of the community. Chapter 5 looks at the role of town centre management and issues of city centre safety, while Chapter 6 discusses the role of the police and of policing in creating safer city centres and has been written by a serving senior police officer and a former senior police officer.

Chapters 7 and 8 look at two specific issues in terms of safety in city centres: public lighting schemes and surveillance by close circuit television cameras. Chapter 9 discusses safer transport to and safer parking provision within city centres. Chapter 10 discusses the provision and design of housing in central areas as a contribution to greater feelings of safety within city centres. Chapter 11 examines the Twenty-Four Hour City concept. Chapter 12 examines some of the issues surrounding safer city centres in the USA. Chapter 13 consists of case study examples of differing approaches to creating safer city centres

(Coventry and Nottingham) where positive actions have been taken to address people's fears and concern about the city centre.

The final chapter summarises the themes and issues raised in the book. It reiterates the limitations of what physical and environmental improvements can achieve and the argument that while crime cannot be 'designed out', environments can be made safer by reducing the opportunities for crime to occur. It reiterates the general thesis of this book that the exigencies of safer environments are not dichotomous with an attractive or rewarding environment, indeed they help maintain the vitality and viability of that environment. Alleviating the fear of crime and making city centres safer in absolute terms is a necessary – although by no means a sufficient – part of their general revitalisation.

1

The Death and Life of City Centres

> The 1960s may have left it looking uglier, but at least it still had a clear purpose. The city then was the centre of social life, the place in which institutions naturally gathered, where ambitious corporations believed they had to have their headquarters . . . They were where we all looked for the kind of public life that gives cities their special quality . . . the chance meetings and random, unexpected social accidents of life. They were characterised by the café and the court house as well as the cinema and the university. The city centre was also the place that could accommodate the awkward, not always picturesque aspects of urban reality that suburbs find too uncomfortable to deal with . . . The changes, social as well as technological, of the 1990s are threatening their very existence. (Sudjic, 1996).

Beyond the necessity of an agricultural surplus, one of the original and most basic reasons for the foundation of settlements was that it was safer within them than without and settlements were often sited for defensive reasons. In recounting the history of the impact of fear on urban design, Chermayeff & Alexander (1963, p.74) note that the normal circumstance for cities has been a climate of fear and insecurity:

> The earlier, coherently organised city was fortified against the invader and the relative anarchy of the larger countryside; the individual buildings and houses were heavily armed against rebels, robbers and the stranger. These precautions, which were clearly expressed in plan and structure, have slowly given way under the influence of economic improvement and social discipline, implemented by the organised enforcement of law.

Even in the early 1960s, they were able to note that the 'great sheets of easily shattered glass, the picture window, the absence of walls and fences' were symbols of 'a short-lived confidence in the efficacy of things . . . [and] . . . opportunities for invaders of the public realm are returning' (Chermayeff & Alexander, 1963, p.74).

The loss of that short-lived confidence has meant a return to a defensive architecture – the 'fortress city' with a high-tech and highly militarised police

Figure 1.1 The Renaissance Centre, Detroit. With its hotel, office and some shopping facilities, the Renaissance Centre is a fortified citadel in a sea of almost total dereliction and abandoned buildings.

force whose 'technologised surveillance and response supplant the traditional patrolman's intimate "folk" knowledge of specific communities' (Davis, 1990, p.251). Over the last two decades, where development in city centres has occurred, it has often tended to be of a defensive nature: gated residential complexes; fortress hotels and office complexes, of which Bunker Hill in Los Angeles and the Renaissance Centre in downtown Detroit (**Figure 1.1**) are obvious examples. At a smaller scale, retail premises in city centres are increasingly protected by steel mesh screens. The screens are mostly pulled down out of shopping hours, but in some city centres they are never pulled up. Davis (1992a, p.7), for example, notes that even prior to the 1991 Los Angeles riots, most liquor stores had completely caged in the area behind the counter, with firearms discreetly hidden, while 'even local greasy spoons were beginning to exchange hamburgers for money through bullet-proof acrylic turnstiles' (Davis, 1992a, p.7–8). Such dystopian visions of the possible future of cities are well described by Mike Davis in his essays, *Fortress Los Angeles: The Militarization of Urban Space* (1990, p.221–263; 1992b) and *The Ecology of Fear* (1992a). Davis also offers a provocative mapping of the 'ecology of fear' (**Figure 1.2**).

The popular aversion to higher levels of taxation and state spending on collective solutions has meant an increasing emphasis on personal safety, private investment in physical security and social insulation among those of a similar socio-economic and/or cultural group. As a result, Michael Sorkin (1992, p.xiii–xiv) observes a characteristic of the emerging city in the US is:

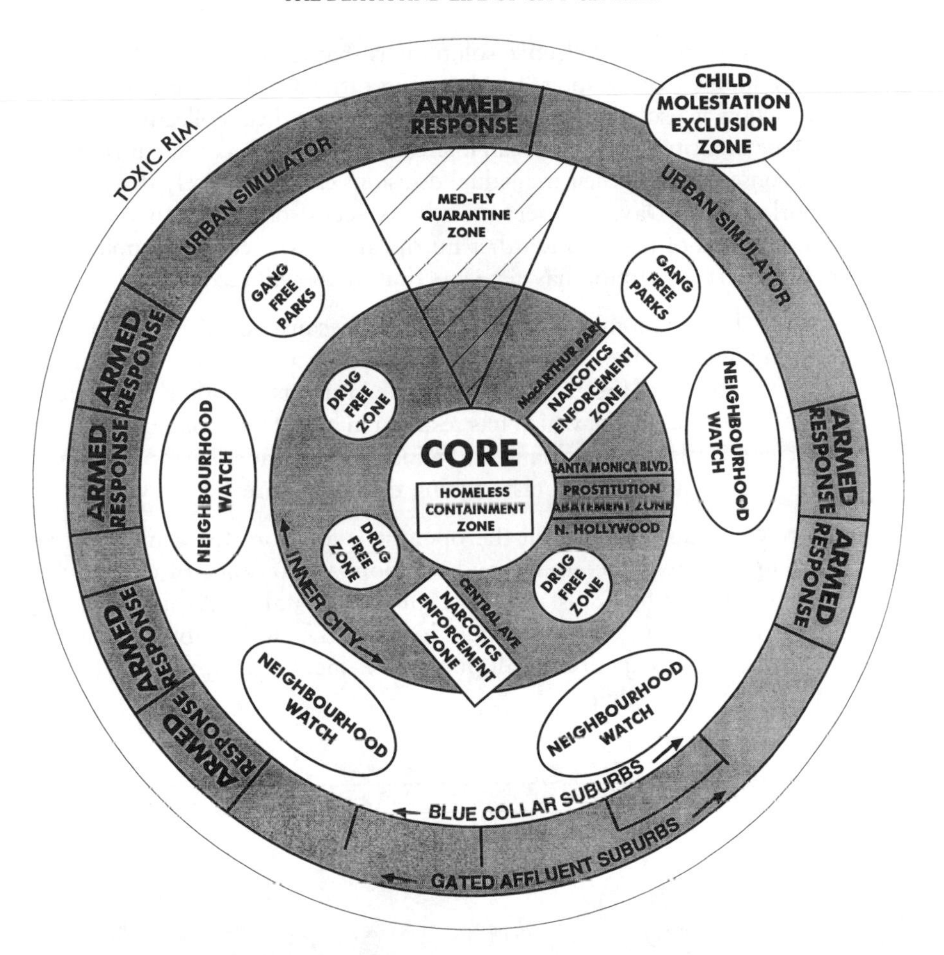

Figure 1.2 The ecology of fear (Source: Davis, 1992a). Davis' map is in the manner of the human ecology school. Taking Burgess 'back to the future', it presumes such 'ecological' determinants as income, land value, class and race, but adds a decisive new factor – fear. Davis (1992a, p.20) asks whether an 'ecology of fear' will become the natural order of the twenty-first century American city?' 'Will razor wire and security cameras someday be as sentimentally redolent of suburban life as white picket fences and dogs named Spot?'

> its obsession with 'security' with rising levels of manipulation and surveillance over its citizenry and with a proliferation of new modes of segregation. The methods are both technological and physical. The former consist of invasive policing technologies . . . and a growing multitude of daily connections to the computer grid . . . The physical means are equally varied: parallel, middle-class suburban cities growing on fringes of old cities abandoned to the poor; enclaved communities for the rich.

The essential problem is that the doubt that collective welfare programmes will work (coupled in any case with a refusal to pay for it) gives more confidence in individual private security. A danger of situations where individuals do not trust or are not prepared to pay for collective solutions is that it may present 'prisoners' dilemma' conditions (see Chapter 4) where the rational self-seeking actions of individuals acting in isolation leads to a reduction in their collective welfare. Although none of a city's citizens has an interest in their city becoming unsafe, the unintended consequence of their individual decisions can be precisely that.

While Sorkin's and Davis's observations may seem somewhat alien and extreme in the context of British and other European cities, Jacobs & Appleyard's (1987, p.114–115) description has elements that many will recognise:

> [Cities] . . . have become privatized, partly because of the consumer society's emphasis on the individual and private sector . . . but escalated greatly by the spread of the automobile. Crime in streets is both a cause and a consequence of this trend, which has resulted in a new form of city: one of closed, defended islands with blank and windowless facades surrounded by wastelands of parking lots and fast-moving traffic. (**Figure 1.3**).

Furthermore, as Sudjic suggests in the opening quotation, for a myriad reasons, the political, social, economic and symbolic significance of city centres has declined and they have lost both viability and vitality. As a result, city centres are losing their role as the critical locus of the city's public realm and there are fears of the extinction of all public life. Jacobs & Appleyard (1987, p.114–115) also observe that:

Figure 1.3 Rear of the Broad Marsh shopping centre, Nottingham. In-centre shopping centres have turned their backs on adjacent streets and internalised street level activities, thereby removing life and vitality from the public realm of the city.

> As public transit systems have declined, the number of places in American cities where people of different social groups actually meet each other has dwindled. The public environment of many American cities has become an empty desert, leaving public life dependent for its survival solely on planned formal occasions, mostly in protected internal locations . . . Fear has led social groups to flee from each other into homogeneous social enclaves . . . It is an alien world for most people. It is little surprise that most withdraw from community involvement to enjoy their own private and limited worlds.

This book discusses making city centres safer. Concern about a lack of safety is one of the factors making a significant contribution to the decline of city centres. As the basic requirements for successful retailing are maximum visibility, accessibility and security, lack of security in city centres means a loss of trade. As the more mobile and higher-income shoppers desert the declining city centre, the area becomes dominated by lower-income and ethnic-minority shoppers, entraining a further downward spiral: higher-income shoppers prefer to use retail areas used by people like themselves (Citizens Crime Commission 1985) and try to avoid 'visiting retail environments . . . frequented by persons considered to be unsavoury or uncongenial' (Beck & Willis, 1995, p.55). This chapter discusses the forces affecting the context for crime and disorder, the city centre, and is in five parts. The first part discusses Modernist urban form and urban planning. The next three discuss the suburbanisation, the disurbanisation and the reurbanisation of cities. The fifth part forms a conclusion to the chapter.

MODERNIST URBAN FORM AND URBAN PLANNING

For the purposes of this book, mainstream urban planning and urban developments from the end of the Second World War to the early 1970s will be regarded as a period of the implementation of Modernist ideas. In both architecture and planning there is a *relative* consensus on what this term represents: Le Corbusier, Charter of Athens, the Bauhaus, and a preference for aesthetics that symbolise the Machine Age and industrial production. Given the conservationist ethos that has prevailed since the early 1970s, much of the more disruptive, larger scale change within city centres occurred prior to the mid-1970s and was largely informed by Modernist ideas of urban space design.

Modernist concepts in architecture and urban planning emerged in the late nineteenth and early twentieth century as a reaction to the physical conditions of the newly industrialised cities of the period with their squalor, congestion and overcrowding. The solution was generally agreed to be more light and air, through a prescription of decongestion and lower residential densities. Given the necessity to separate dirty, noxious heavy industries of the period from residential development, functional zoning also had a strong appeal and appeared in the earliest Modernist city planning ideas of, for example, Soria y Mata's *La Ciudad Lineal* (1882) and Tony Garnier's *Cité Industrielle* (1898). In the twentieth century, zoning was also a fundamental part of the International Congress of Modern Architecture (CIAM) – an architectural forum established in 1920s by leading Modernist architects. The report of its 1933 Congress and the group's best known treatise, *The Charter of Athens*, prescribed a rigid functional zoning of city plans with green belts between areas reserved for different functional uses. The rationale of functional zoning was not just the environmental benefits but also, it was argued, because it would provide a more economically efficient and rational city structure.

Modernism was also a response to the perceived challenges and opportunities of the modern period. It both heralded and drew stimulus from the perception of the start of a new age – the Machine Age – the age of industrial production. Accordingly, Modernists had a strong belief in progress and in the technological potential of the new age. Prior to the Modern period, change in the urban fabric had generally been slow, enabling successive generations to derive a sense of continuity and stability from their physical surroundings. Modernism and modern development changed this: the historic city form was seen as obsolete, a hindrance to the future, and an obstruction to attempts to build in a rational and functional manner to exploit the opportunities and relieve the problems of the modern city. Hence comprehensive redevelopment of large sites was favoured over a more incremental rehabilitation and refurbishment of buildings and areas.

Redevelopment also permitted significant physical improvements over the earlier urban form. Architects and planners had become interested in research to determine the optimum design and distribution of building blocks to maximise the access to sun and daylight, and minimise overshadowing. As a result higher rise slabs and point blocks were favoured. The ambition was to create environments that provided healthier conditions than the slums of the industrial cities, and buildings that would provide healthier internal conditions. The best way to obtain more light and air was to build vertically, and the building form that readily exploited the new modern construction techniques and materials was the skyscraper. Skyscrapers offered a totally new image to the city and the potential future of urban areas.

The result of the interplay of the various factors was a new concept of urban space and urban form. As Summerson (1949, p.191) said of Le Corbusier's *Contemporary City*: 'The park is not in the city, the city is in the park'. Traditional, relatively low rise streets and squares were eschewed in favour of a rational – usually orthogonal – distribution of slab and point blocks, set in parkland or urban plazas. Such ideas changed the nature of urban space design and lead to an erosion of the traditional, spatially defined urban public realm. 'Streets' were rejected in favour of 'roads', with roads being seen exclusively as channels for movement rather than having the social qualities or focus that traditional streets possessed. In the post war British New Towns, even the relatively benign concept of the neighbourhood unit (see for example Osborn & Whittick, 1977) often had separate vehicular roads and pedestrian routes rather than traditional streets.

Combined with this concern for healthier internal conditions was a concern for the aesthetic expression of the building's function. Rather than being the edges which defined public spaces, buildings became freestanding sculptures or objects in space following *primarily* their own internal logic rather than responding *primarily* to their urban context. With an increasing agglomeration of such buildings, a city gradually loses its spatial coherence becoming a 'jumble' of competing or isolated monuments and small complexes of buildings surrounded by roads and a sea of car parking.

SUBURBANISATION

Modernist ideas of urban form and urban planning sought to create better cities and urban environments than would be achieved by the unfettered play of market forces. Nevertheless, the twentieth century has seen powerful forces of change affecting the morphological evolution of cities. Modernism was just one of the forces of change. The aggregate effect of such forces is a tendency towards one of two outcomes – either *dispersal* or *concentration*. As Knox (1987, p.329) notes, the net outcome in any particular place depends on the relative strength and balance of these forces and varies from place to place according to economic circumstances and historical antecedents.

In the twentieth century, as a result principally of increased mobility (both physical and electronic), social aspirations and subsequent economic choices by those with effective demand, the predominant trend – in the developed world at least – has been towards dispersal. Pressman (1985, from Knox, 1987, p.329) lists the following trends which tend to create a more dispersed urban pattern: the increasing capacity and sophistication of electronic communications; the increasing freedom from the locational constraints affecting all kinds of economic activities; the suburbanization of 'back-office' white-collar jobs; increasing preferences for lower-density, rural character environments; the increasing congestion and rising operating costs of older public transport systems; and the internationalization of the economy.

Following the urbanisation precipitated by the Industrial Revolution, when the main means of transport had been by horse or by foot, the introduction of mass transport systems broke the necessarily close spatial relation between workplace and residence. After about 1870, many cities began to develop suburban railways. Further developments in the early 1900s saw horse-drawn and then motorised trams and buses. This was followed in the largest cities in the years before and after the First World War by underground railways. The initial motivation for suburban residential locations was to escape the central city's negative associations, pollution and crime, but there were also strong attractions: better quality of housing, the possibility of gardens, healthier living conditions and the social status such locations conferred. In the 1930s, the steady salaried employment of the burgeoning middle classes enabled banks to lend money as mortgages. This fuelled speculative development which, together with the expansion of transport systems and – in most countries – an ineffective planning system, led to further suburbanisation.

The post-World War Two period saw further suburbanisation through the widespread individual mobility afforded by increasing car ownership. Increasing distances that people were either willing or forced to commute continued this dispersal. The social aspirations and economic choices of the period were also reinforced by planning and urban design theories and professional attitudes. By the 1970s and 1980s, the suburban impulse was further reinforced by a new factor – fear.

Changes in the Central City

In the post war period while extensive suburbanisation was happening, major physical changes were happening within the central city. Facing the challenge of building a better future after the Second World War, post-war governments drew upon the wartime experience of mass production and planning as a means to undertake major reconstruction programmes. An important part of this was the re-building, re-shaping, and renewal of the urban fabric and the informing context for this was Modernism and the ideas of CIAM.

As the physical improvements offered by redevelopment were generally desirable, reconstruction programmes were complemented by slum and city centre clearance and redevelopment programmes. The post war approach was codified in the 1947 *Advisory Handbook for the Redevelopment of Central Areas* (Ministry of Town & Country Planning, 1947). This document emphasised open planning (creating formless 'amenity space' through the application of daylight indicators and floorspaces indices) and vehicle circulation (free flow road systems and linked parking provision) (Punter, 1990, p.9). The Handbook also emphasised the segregation of functions through zoning (particularly the zoning out of residential uses, unmixing retail and office uses, and confining institutions to campuses). Functional zoning was viable because new forms of transport, including the motor car and the prospect of mass car ownership, would be able to tie the separated areas together. As was evident in Le Corbusier's lovingly rendered images of roads full of cars, the car was a potent symbol of the potential drama, excitement and freedom of the modern age.

The major cause of change in the urban environment in the postwar period was increasing car ownership. Planners and architects argued that existing cities were ill-equipped for motor cars and other forms of mechanised transport and would have to be radically restructured. In the 1950s and 1960s, as an increasingly affluent population moved further into the suburbs, inner area road building schemes were proposed to relieve congestion and meet the increasing need for access from the suburbs to and around the city centre. The introduction of sufficiently powerful planning legislation permitted urban motorway schemes and other transport improvements through existing towns. Thus, clearance policies were augmented by grand road building schemes intended to restructure cities to ensure efficient access for cars (see for example Buchanan, 1963).

The Reaction to Modernism

Although comprehensive redevelopment and large road building schemes could be seen as necessary for the city's economic salvation, by the mid-1960s the social effects of post-war urban planning policies were becoming evident with increasingly widespread public protest at clearance, redevelopment and road building plans. Modernist ideas of urban form and urban planning – albeit well-intentioned – were proving highly unsatisfactory in practice. As Birch & Roby (1984, p.200) argue, books such as Jane Jacobs' *The Death and Life of Great American Cities* (1961) and Herbert Gans' *The Urban Villagers* (1962) cautioned professionals to be: 'more aware of the diverse, smaller-scale building blocks of planning and more appreciative of the beauty and functionalism of existing neighbourhood organization'.

Large scale redevelopments – in America, termed the 'Federal Bulldozer' – inevitably disrupted historic patterns of settlement and communication. The physical disruption caused by large-scale redevelopment and by road building schemes, combined with the design of the new developments, destroyed historic street patterns and traditional notions of urban space. Furthermore, while the process of redevelopment was highly disruptive to small firms and businesses, the product was also fatally flawed: large, relatively simple blocks inevitably simplified the land use pattern, removing the 'nooks and crannies' that could house the economically marginal but socially desirable uses and activities that gave variety and life to an area.

A loss of urban vitality also resulted from a rigid, planned functional zoning of cities and urban areas as British and many European cities started cleansing their city centres of uses deemed incompatible with an image of modernity. In Britain, the prototype for the post-war city centre was Coventry, which was completely redesigned and rebuilt after extensive bomb damage (see also Chapter 13). The logic of zoning was reinforced by transport developments and high land values which tended to push out lower value uses, resulting in an erosion of the complexity and vitality of city centres, and shaping them into islands of activity alive during the day only. The tendency towards sterility was further exacerbated by large, generally monofunctional office blocks whose limited number of entrances internalised much of the traditional street life and activity of the city.

There were – and have been – many critics of a rigid zoning approach. Jacobs (1961, p.130) argued that a great part of the success of city neighbourhoods depended on the overlapping and interweaving of activities and areas and that urban areas were traditionally characterised by their mix of uses. Furthermore she argued that a range of different building types and ages, with their variety of renting profiles, was vital to the life of urban areas (Jacobs, 1961, p.200). Alexander (1965) in his seminal essay *A City is Not A Tree*, argued the merits of a 'semi-lattice' over a 'tree' structure for a city: tree-like structures led to rigid separations but semi-lattices contained complex overlappings and mergings.

DISURBANISATION

Initially suburbanisation was largely limited to residential, industrial and related functions, but from the 1950s and 1960s onwards there was increasing decentralisation of other functions, especially of retail and, later, of office functions. Initially the first retail function to decentralise was convenience shopping, especially food. This was followed by those selling bulky items that were generally difficult to transport or required large display areas, such as furniture and carpets. Subsequently there was a general movement to out-of-town locations by other traditional city centre functions, often because they had a symbiotic or dependent relation with the retail – once the retail had gone they became locationally obsolete. What was therefore happening was a functional disurbanisation. In terms of accessibility, two principal factors accounted for this. First, the development of motorway systems and ring roads made the edge of cities relatively more accessible, and secondly, the inner urban roads had either not been implemented or were unable to relieve the central city's traffic congestion sufficiently.

The Loss of Centrality and the Edge City

Improvements to the road system originally intended to save the city centre (see for example Buchanan, 1963) were superseded by further developments. The intersection of ring-roads with a hub and spoke road form creates a grid with peak land values occurring near to the most frequented intersections. Evidence of this, as Garreau (1991) notes, is a morning traffic jam in *both* directions. Thus, the traditional model of the city – where land prices decline with increasing distance from the city centre as they are traded off against transport costs deemed to increase with distance from the city centre – is now obsolete: a few minutes in central Detroit serves to illustrate the reality of this observation (**Figure 1.4**).

Central city decline was perhaps inevitable once it ceased to be the pre-eminent hub and focus of the transport system. As Fishman (1987) notes this had already been foreseen separately by H G Wells and Frank Lloyd Wright. Wright for example argued that 'the great city was no longer modern' and was destined to be replaced by a decentralised society. Wright called his new society *Broadacre City*. *Broadacre City* was based on universal car ownership combined with a network of super highways, thereby removing the need for population to cluster in a particular spot: 'indeed, any such clustering was necessarily inefficient, a point of congestion rather than of communication' (Fishman, 1987, p.188). Although often confused with a universal suburbanisation, Wright's concept was the opposite of suburbia: suburbia represents the extension of the city into the countryside, *Broadacre City* represents the disappearance of all previously existing cities (Fishman, 1987, p.188).

The last two decades have seen the emergence of what Joel Garreau (1991) has termed the 'Edge City'. According to Garreau (1991, p.4):

> Edge Cities represent the third wave of our lives pushing into new frontiers. First, we moved our homes out past the traditional idea of what constituted a city . . . Then we wearied of returning down town for the necessities of life, so we moved our market places out to where we

Figure 1.4 Central Detroit. Central Detroit is a classic example of central city decline. Some of the vacant buildings have been cleared and, in lieu of redevelopment, trees and grass have been planted as a desperate means of beautification.

> lived . . . Today, we have moved our means of creating wealth, the essence of urbanism – our jobs – out to where most of us have lived and shopped for two generations. That has led to the rise of the Edge City.

Interestingly, Garreau (1991, p.113) also links the growth of edge cities to increasing car ownership by women and their increased participation in the labour force. Nevertheless, research in Nottingham (Crewe & Hall-Taylor, 1991) has shown *inter alia* that those women who do not have access to cars generally prefer employment in or near the city centre. Such locations usually only require a single bus ride from home to work and allow access for other activities such as lunchtime shopping.

The emergence of edge cities can be seen in commuting patterns. In America, for example, the traditional commute from suburb to downtown is increasingly a minority pattern as most of the trips taken by metropolitan Americans skirt the old centre (Pisarski, 1987). Fishman (1987, p.17) also observes how 'the simultaneous movement of housing, industry, and commercial development to the outskirts has created perimeter cities that are functionally independent of the urban core'. The phenomenon has been most prominent in the USA, but in Britain especially and in Europe more generally evidence of its emergence is becoming apparent. In Britain, the urban form which is increasingly emerging is the 'doughnut' city: a declining central city with a new linear – or 'edge' – city wrapped around its periphery.

Out of Centre Shopping Malls

One of the most instrumental factors in the decline of city centres was the decentralisation of retail functions. As a consequence of both the changing transport system and the increasing number of people abandoning the central city as a place to live, developers quickly saw that rather than bringing the people to the shops, it would be more profitable to take the shops to the people. In America, from the 1950s onwards, suburban shopping malls became a constituent feature of the 'American Way of Life' (Poole, 1991). In 1950, there were an estimated one hundred such centres in the US. By the late 1970s, there were over 20,000. Covered shopping malls clustered all shopping needs into one purpose-built location, often in a protected and climate-controlled environment. The extremities of the North American climate – extreme heat in the south and extreme cold in the north – made covered malls with immediately accessible parking even more attractive to shoppers than the outdoor downtown shopping areas. North America also saw the emergence of 'the strip' as a location for low rent activities. Strips were designed to accommodate car access: while people might still walk on streets, nobody walked the strip.

City Centre Shopping Malls

In the late 1960s and through the 1970s and 1980s, in an attempt to staunch the haemorrhage of retail investment from the city centre, many local authorities built city centre shopping malls to rival those being built outside the city. Attracted by the vision of American-style shopping, the first two covered malls in Britain were opened in 1964: one at the Elephant and Castle in London, the other in Birmingham (Poole, 1991). In-centre shopping malls were an attempt to both emulate the advantages of out-of-town shopping malls and to address some of the problems of city centres. The shopping mall's 'total environment' gives 'the public good reason for feeling safer there than on downtown streets. Malls have better lighting, a steadier flow of people, and fewer hiding places and escape routes for muggers' (Frieden & Sagalyn, 1989, p.234).

While bolstering the retail function in city centres, the construction of large suburban-type shopping plazas with multi-storey car parks fostered further reasons for the general downward spiral of city centres. Functionally many in-centre shopping malls turned their back on adjacent streets, except at key points of entry into the peak pedestrian flows, and internalised street level activities. When originally built, in addition to providing shops and restaurants, many malls were intended to be thoroughfares and public spaces. The malls, however, rapidly became highly-controlled enclaves for shoppers with cars and access was restricted out of shopping hours.

With restricted access, the malls destroyed the grain of the townscape and reduced the permeability of the town centre. Thus, as British city centres typically close at 5.30 pm, city centre shopping malls mean that much of the heart of the city centre is also locked up at night creating large dead spaces. The population, closeted away in the suburbs, fear urban areas after nightfall and do not return to them, seeing them as dark, lonely places, ill-served by public transport and dominated by those seeking criminal opportunity (Poole, 1991). Since the late 1960s, this perception has been reinforced by a general dissatisfaction with the products of Modernist architecture, urban design and city planning, such as the destruction of existing historic and traditional urban environments and the building of isolated structures surrounded by open space. Within the city centre itself, there is competition for customers and retailers between the shopping mall and the High Street and traditional city centre retail locations. Thus, in the face of the wider trends, city centres did not necessarily help themselves.

The Decentralisation of Retail

Despite the development of city centre shopping malls, retail decentralisation has continued. Schiller (1988) identifies three distinct phases or 'waves' of the decentralisation of retailing that have taken place in Britain. The first wave took place during the 1970s, when a major change took place in food retailing as supermarket chains opened stores located on the edge of urban areas. Such stores provided convenience goods under one roof, free and spacious car parking in locations easily accessible by car. The effect on city centres was limited as it only removed comparison food shopping; city centres continued to be the focus for the non-food (comparison goods) sector.

Schiller's second phase occurred in the 1980s and consisted of the movement of bulky goods, such as DIY, electrical goods, carpets and furniture from town centres to retail warehouse parks on the periphery of urban areas. This form of retailing demanded larger floorspace areas and sites in accessible locations which city centres were increasingly unable to provide. As such retailing outlets met a specific demand and sold a relatively limited range of goods, they were not viewed as being in direct competition with city centres.

Schiller's third phase occurred from the mid-1980s onwards and involved the development of regional shopping centres and retail parks. These have had the greatest impact on city centres as they involve the retailing of high order goods which are in direct competition to those goods sold in town centres. Regional shopping centres are usually defined as those with more than 50,000 square metres (540,000 square feet) of retail and leisure space, away from the centre of a town. Retail Parks developed from 1984 onwards when Marks & Spencer announced that they would be implementing a programme of out-of-centre store openings.

In the mid-1970s there were fewer than twenty out-of-town shopping complexes in the UK. There are now nearly eight hundred and 65% of the population is within 30 minutes drive of one of the five existing regional shopping centres: Brent Cross, North London; Lakeside, near Thurrock (1.2 million sq ft); Meadowhall, near Sheffield (1.1 million sq ft); Merry Hill, West Midlands (1.4 million sq ft); MetroCentre, near Gateshead (1.6 million sq ft) (**Figure 1.5**). This proliferation of out-of-centre complexes permits the mobile shopper choice. There is the appeal of the shopping environment: high quality, newly-built, environmentally-controlled shopping malls

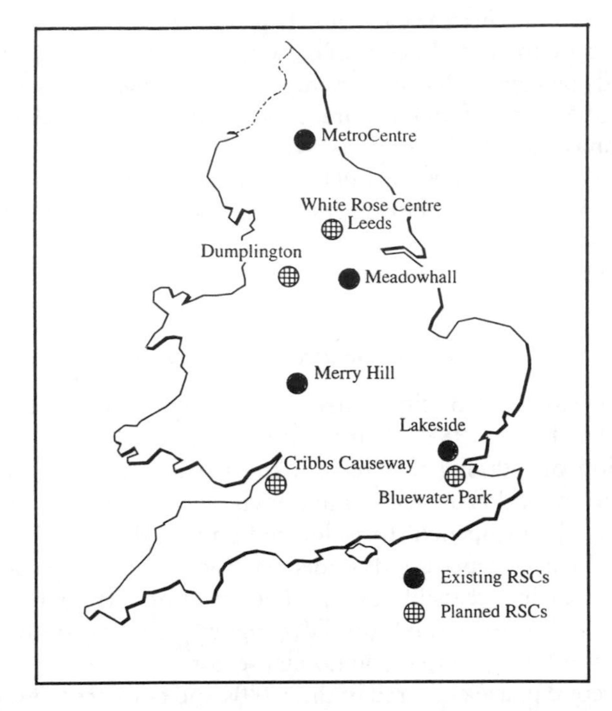

Figure 1.5 Locations of existing and planned regional shopping centres in England.

providing comfort and security, while their single management permits a uniform maintenance of standards and quality. In addition, as ease of travel and parking figure strongly in the decision of where to shop, key attractions of out-of-centre shopping complexes are their good accessibility and safe, convenient – often free – parking.

Central City Decline

Decline in town centres is initially gradual then rapidly spirals downwards: major stores and magnets move out, downmarket and discount stores move in, buildings fall into disrepair as rents are driven downwards and eventually there is physical dereliction. The decline of city centre retail is most marked in the United States. The *Revitalising Downtown* survey by the National Trust for Historic Preservation (1988) of a sample of American towns of 10,000 people or more showed that in 1976 70% had their prime retail locations in their downtown areas. By 1986, that figure had declined to a mere 26%. The spectre looms of European cities following North American trends.

Sheffield city centre has lost 15–20% of its trade through the opening in the late 1980s of the nearby Meadowhall regional shopping complex. Nottingham, much further away, has lost 5%. Birmingham city centre lost up to 10% of its trade due to the opening of Merry Hill regional shopping centre (Hobson, 1995). In smaller towns the effect can be more dramatic. Research by

Roger Tym (1994) showed that Dudley had lost 70% of its trade after the opening of Merry Hill. Beck & Willis (1994, p.23) note research which suggests that only the top 200 high streets in the UK are 'doing well' and 'in many areas there is a steady drift of customers towards larger city centres and shopping malls'. Thus, as more out-of-town shopping centres are built drawing activity and life from central areas, problems of vacant shops and retail premises, deserted city centres and urban streets are beginning to emerge as a major problem in all developed countries.

The Loss of a Sense of Place

As the traditional centre loses the benefits of functional agglomeration and its primacy in the transport network, what happens is the dissolution of the central city as a political, economic, social, and symbolic locus. Fishman (1987, p.198) gives the planners' criticisms of this as a social and economic disaster which has 'segregated American society into an affluent outer city and an indigent inner city' and erected 'even higher barriers that prevent the poor from sharing in the jobs and housing of the technoburbs'. Dispersal is also seen as a cultural disaster: while the rich and diverse architectural heritage of cities decays, the wealth generated by post-industrial America 'has been used to create an ugly and wasteful pseudo-city, too spread out to be efficient, too superficial to create a true culture'.

One of the consequences of this loss of centrality has been the concomitant cultural loss of a sense of place and location. As a consequence of these trends, the central city and the city centre have become less significant as a focus of people's lives; people can live, work and play without ever visiting the city centre. With increasing impact, various technological developments in the post-war period have enabled an increasingly domesticated leisure. Fishman (1987, p.185) observes that the new city's true centre is not the downtown business district but each individual residential unit: 'From that central starting point, the members of the household create their own city from the multitude of destinations that are within suitable driving distance'.

For perhaps the first time ever, people have had the option to reject the central city, and many have taken this option – or, as a result of fear, they have chosen not to take the risk. Yet precisely because in the edge city all places are alike, it is the central city which is synonymous with and symbolically represents the city. The concern of this book is with cities which still have some form of recognisable city centre in the traditional pattern to serve as the focus for revitalisation. It is therefore applicable to most European cities and cities elsewhere in the world, but a more limited number of North American cities. For those cities which have become fully ageographic it is perhaps already too late.

In Britain, the most recent planning guidance – *PPG 6: Town Centres and Retail Developments* (DoE, 1996a) – is very supportive of city centres. This and an earlier PPG 6 (DoE, 1993) mark a distinct sea change in planning policy from the position of the late 1980s (for example DoE, 1988) which had been firmly committed to a non-interventionist, *laissez-faire* approach. The new policy guidance is more protectionist seeking to remove the element of competition posed by out-of-centre developments by imposing restraint policies to hamper or prevent their development. Those seeking to build out-of-centre shopping complexes now have to prove that no areas are available in or on the edge of city centres. This may, however, simply be a case of shutting the stable door after the horse has bolted. Four further regional shopping complexes are planned and are likely to obtain planning permission: Trafford Centre, Dumplington, Greater Manchester (1 million sq ft); Cribbs Causeway, near Bristol (0.7 million sq ft); the White Rose Centre, near Leeds (0.65 million sq ft); and Bluewater Park, Kent (**Figure 1.5**). The latter is linked with a Channel Tunnel station making it less than two hours' travel time from northern Germany. The Department of the Environment also commissioned the *Vital and Viable Town Centres: Meeting the Challenge* study (URBED, *et al.*, 1994) to catalogue examples of good practice.

REURBANISATION

City centre decline is exacerbated by a time lag as property owners adjust to new market conditions caused by changes in the underlying pattern of accessibility and demand and their continued expectation of – and hope for – premium rents. Once property owners have adjusted to new market conditions by lowering their rents, the decline of city centres as the prime business location presents an opportunity. Falling land values and rents make them ripe for redevelopment or conversion for other activities. Thus, the partial loss of the retail function and other functions can be seen as an opportunity for new developments and functions within the city centre.

Although retailing helps to stabilise and anchor their local economies, shopping is not the be-all-and-end-all of city centres: they are 'city centres' not 'shopping centres'. Here city centres have a major advantage over out-of-centre shopping complexes. They already have mixed uses: shopping, restaurants and cafés, entertainment, leisure and tourism activities of varying kinds. They also have established character and historicity. The challenge is to find a new economic role or niche for the city centre. This has often involved new uses: tourism, services, the arts, and housing. It is not only the real estate value that developers seek to recoup, but also the 'cultural capital' represented by city centres and downtowns (Lovatt & O'Connor, 1995, p.127–128).

The 1980s and early 1990s have seen localised successes of city centres and waterfronts revitalised by tourism and entertainment that have resisted the trend towards decentralisation. By so doing, such projects have re-invented a positive notion of 'urbanity'. Among the earliest and most publicised examples were Quincy Market, Boston, and London's Covent Garden (**Figure 1.6**). The re-urbanisation however has been partial and piecemeal. Rather than a comprehensive revitalisation of the central city, certain sectors and enclaves have been revitalised. Rather than a generally safe city with a few unsafe spots, it has become – especially in America – a generally unsafe central city with certain well-controlled and regulated safe area or spots.

Figure 1.6 Quincy Market, Boston. From the late 1970s onwards, some projects have resisted the trend towards decentralisation and have revived a positive notion of urbanity.

The New Urbanity

The positive reception to the regeneration of parts of the central city has also seen a desire for more 'urbanity' in new developments. David Harvey (1989, p.68) for example describes how one theorist, Leon Krier, 'seeks the active restoration and re-creation of traditional "classical" urban values'. This is interpreted as meaning either the restoration of an older urban fabric and its rehabilitation for new uses, or the creation of new spaces that express the traditional visions. In architecture, urban design and planning, Modernism has given over to postmodernism. In this book, postmodern refers to the historical condition or period of being after Modernism.

Postmodernism in urban space design has been principally informed by two key themes both of which – while somewhat conservative – hold a positive view of urbanity aiming to re-establish 'traditional' urban values. The first theme entails a desire to keep existing and familiar environments intact. By the late 1960s, the value and social characteristics of traditional environments were increasingly being recognised in contrast to – and as a reaction against – Modernist environments. The conservationist ethos was also part of a more general shift away from the idea of creating cities *de novo* in favour of stressing the sense of place and historical continuity (Knox, 1987, p.340), and, in particular, a recognition of the qualities of the existing built environment, its cultural, social and historic attributes.

The second theme entails a greater appreciation of traditional urban processes and precedents and a new consideration of the qualities and scale of the 'traditional' city. For influence many theorists and practitioners have returned to – often romanticised – images of the period of urban evolution prior to the Industrial Revolution. Many critics, however, regard the new urban realm which draws explicit reference from – or even copies – historical precedent as 'false' and 'lacking' authenticity. In Michael Sorkin's (1992, p.xiv) words, it is merely:

> a city of simulations, television city, the city as theme park. This is nowhere more visible than in its architecture, in buildings that rely for their authority on images drawn from history, from a spuriously appropriated past that substitutes for a more exigent and examined present.

While the influence can to varying degrees be 'ironic' or 'earnest', it nevertheless represents a renewed desire – however partial and/or limited – for the physical definition of urban spaces and for urbanity in general. The postmodern approach therefore often seeks to build a physical public realm which emulates the streets and squares of traditional, pre-industrial cities.

The New Urban Socio-Cultural Public Realm

Although much of the contemporary urban design debate is concerned with the physical significance of traditional urban spatial arrangements and urban forms, it is more than the spatial definition of the public realm which is important: the public realm is both a physical *and* a socio-cultural construct. Not only is a spatially defined physical public realm required, that public realm needs to be naturally animated by people: spaces become places through their use by people. Nevertheless, concurrent with the decline of a spatially articulated urban public realm has been the relative decline of the socio-cultural public realm. As people have chosen not to use – or have been deterred from using – the city centre, then, as Ellin (1996, p.149) describes, activities which had previously occurred in the public realm were either abandoned or 'usurped by more private realms' as leisure, entertainment, gaining information, and consumption are increasingly satisfied at home through the television or computer. Even if one left the home, such activities now often take place in 'the strictly controlled uni-functional settings of the shopping mall, theme park, or variants thereof' (Ellin, 1996, p.149).

Although developments designed in a postmodern manner often include new public spaces, entry and behaviour within them can be tightly restricted. Extending the logic of the highly-controlled environment of the shopping mall, to protect their investment and to ensure safety for their customers, private agencies have sought control over ostensibly public spaces, thereby rendering them increasingly private and exclusive. This, as Sudjic (1996) observes, represents 'the conversion of the city into a playground for those affluent enough to afford its attractions'. As a result, some see the 'new urbanity' as a transmogrified 'urbanity'. Davis (1992, p.16), for example, sees it as largely a 'social fantasy' which 'is increasingly embodied in simulacral landscapes – theme parks, "historic" districts and malls – that are partitioned off from the rest of the metropolis'. More positively, however, Lovatt and O'Connor (1995, p.128) argue that 'however superficial and spatially circumscribed . . . the emphasis on play, strolling and idle socialising could have wider effects'.

MAKING CITY CENTRES SAFER

Despite the more pessimistic prognoses reviewed here, this book is written from the view that cities and their centres will continue to be important. The changes do not necessarily signal the end of cities as significant economic places. In the midst of social, economic and political change, the physical infrastructure and fabric of cities clearly represents a relatively fixed point. There is a lot of committed capital – both real and symbolic – in city centres. Although the impact has been uneven, the economic robustness and the established historical advantages of cities and their central areas are not easily discarded: the transformation of British and European cities into a loose mosaic of edge cities is not a foregone conclusion.

Castells & Hall (1994, p.7) see a positive role for cities and the public and private institutions which make up that city. For them, the most fascinating paradox is that 'in a world economy whose productive infrastructure is made up of information flows, cities and regions are increasingly becoming critical agents of economic development'. Their argument is that as the economy is global, national governments have increasingly fewer powers to act upon the critical processes that shape their economies and societies. Cities and regions, however, are more flexible and can adapt to changing conditions of markets, technology, and culture. Although, they have less power than national governments, they have a greater capacity to generate targeted development projects, negotiate with multi-national firms, foster the growth of small and medium endogenous firms, and create attractive conditions for new sources of wealth, power, and prestige (Castells & Hall, 1994, p.7). How long this will continue, however, is subject to considerable debate. Furthermore, what Castells & Hall refer to as 'the city' is really the city as a whole. The concern in this book is with one particular part of that city – the city centre. A city's economic success should not mask the fact that the success may be unevenly distributed within the city itself.

Pressman (from Knox, 1987, p.329) identifies the following trends that provide a resistance to the tendency towards dispersal.

- the continuing need for face-to-face interaction in certain economic activities
- increasing energy costs and uncertain energy supplies
- the emergence of an anti-technology attitude
- the persistence of an impoverished urban underclass
- the increasing attractiveness of revitalised inner urban settings to some elements of the young middle classes
- the inaccessibility of sophisticated (and expensive) telecommunications to the less-affluent.

What is problematic however is that the concentrating trends do not seem as convincing or as pervasive as those that encourage dispersal. There are some glimmers of hope. Electronic communication is perhaps the most powerful decentralising force, particularly as developments in technology increasingly permit the creative cross-fertilisation and the transferral of the 'ambiguous information', thereby lessening the need for face-to-face interaction (see for example Garreau, 1991, p.134). Yet at present – and as an extreme example – businessmen still find it necessary to fly by Concorde from London to New York and back in a single day to attend business meetings in person. Furthermore, Manuel Castells (1989, p.1) argues that Alvin Toffler's prophecy of 'electronic cottages' is an example of technological determinism and that the application of technology is usually mediated by social factors. Castells notes how various prophecies suggest that, in the informational age, the need for the concentration of population in cities has been superseded, and yet none of these stands up 'to the most elementary confrontation with actual observation of social trends . . . Intensely urban Paris is the success story for the use of home-based telematic systems' (Castells, 1989, p.1–2).

Concerns about environmental sustainability and the use of scarce fossil fuel supplies may ultimately lead to more compact and centralised urban forms. But even then, there is considerable debate and doubt about such predictions. The logic of the 'compact city' is challenged because jobs and retail have already generally moved closer to – and not further from – where people live, with commuting times falling rather than increasing (Pisarski, 1987; see also for example Breheny, 1993; Jenks *et al.*, 1996).

Perhaps the strongest argument for a new centrality is that people feel that something is missing in the placelessness of suburbia and the edge city, in shopping complexes and theme parks. Thus, they crave the excitement, variety, historicity, spectacle and carnival of real cities. As a result, they may choose to patronise the city centre if it still exists as a meaningful option. Solesbury (1993) identifies an emerging ideology which:

> offers a positive view of the potential of cities in place of the negative view of urban problems that has prevailed hitherto. In cultural terms it recognises not just that cities are the depositories of much 'high' culture in their galleries, theatres, museums, concert halls and in their buildings and townscapes, but also that they provide a fertile milieu for a vibrant 'low' culture of fashion, journalism, entertainment, sport and leisure.

Nevertheless, if the city centre's attractions are undermined by crime, the fear of crime and the perception that it is unsafe, then the decentralising forces outlined will be stronger. As a prerequisite, therefore, city centres must be perceived to be safer and, indeed, must be safer. It is this issue which the remainder of this book discusses.

2

The Dimensions of Crime

A full understanding of crime – and any preventative measures – must synthesize and integrate all the dimensions of the criminal event. Brantingham & Brantingham (1991, p.2) argue that 'criminal events can be understood as confluences of offenders, victims or criminal targets, and laws in specific setting at specific times and places'. A full crime analysis, therefore, has four principal dimensions: a 'victim' or 'target' dimension, an 'offender' or 'motivation' dimension, a 'spatio-temporal', 'locational' or 'opportunity' dimension and a 'legal' dimension. The first three are the minimal elements of victim crime. The fourth provides a law to actually define it as a crime, the means to prosecute it as such and, if convicted, levy appropriate sanctions on the offender.

Although this book concentrates on crime and safety in a specific location – city centres – the various dimensions of crime must be discussed to provide the context in which strategies for crime prevention and reduction can be considered. Actions to ameliorate the impact of crime and the fear of crime can be focused at any or several of the various dimensions. The legal definition of acceptable behaviour can be modified, for example, by the introduction of byelaws outlawing the public consumption of alcohol; potential offenders may be demotivated by the threat of punishment or by gainful employment; victims may avoid putting themselves at risk; and the location may be modified to make it more difficult for a crime to be committed there. As the various dimensions also interact with each other, crime and fear of crime reduction strategies need holistic consideration.

THE LEGAL DIMENSION

Crime

The *Oxford English Dictionary* states that a crime is: 'An act punishable by law, as being forbidden by statute or injurious to the public welfare . . . An evil or injurious act; an offence, a sin; esp. of a grave character.' In its broadest sense, however, a crime is committed wherever there is demonstrable harm to a person and/or property. Gottfredson & Hirschi (1990) for example define 'crime' as an act of force or fraud undertaken in pursuit of self-interest. Muncie (1996) identifies a number of different conceptions of crime:

- crime as criminal law violation – ie, a 'black letter law' definition of crime.
- crime as norm infraction – ie, violations of moral and social codes.
- crime as social construct – ie, crime is dependent on *how* certain acts are labelled and on *who* has the power to label (Scraton & Chadwick, 1991, p.289).
- crime as ideological censure – ie, law supplies to some people or groups both the means and *the authority* to criminalise the behaviour of others (Muncie, 1996, p.15).
- crime as historical invention – ie, law (and thereby crime) is created and applied by those who have the power to translate their interests into public policy (Muncie, 1996, p.18).

In expansion of the second and third points, Muncie (1996, p.18–19) observes that the various conceptions of crime can be generated from competing accounts of the social order. If that order is consensual then 'society' can be seen as creating 'rules' and thus, 'crime' is the infraction of that society's legal, moral or conduct norms. When the social order is considered pluralist or conflict-based then the concern is with a 'state' that has the power to criminalise. In this instance, 'crime' refers not to particular behaviours, but to social and political processes whereby those actions are subjected to criminalisation. The distinction is important and can explain different interpretations of criminal behaviour and attitudes to methods of crime prevention and control.

In this book, the concern is with the infraction of legal, moral and conduct norms, rather than the more politically motivated or justified crime. It is recognised however that crime is often a social construct, and, thus, as the society's social and moral codes define in a general sense what is – and what is not – considered to be 'acceptable behaviour', the definition of a 'crime' will change from society to society. Nevertheless, it is the legal framework which effectively establishes and defines the 'acceptable' parameters of behaviour within a society.

The legitimacy of government (both local and central) and/or other public authorities to impose restrictions on the behaviour and freedoms of its citizens in the interests of the wider public good inevitably raises issues of civil liberties and of individual freedoms. In the UK, people do not have express rights, rather they have residual rights – those which the state has not expressly taken away. In other countries, citizens do have express rights – such as the 'right to bear arms' – which are guaranteed under the constitution.

Freedom is difficult to define precisely. Nevertheless, Lukes (1985, p.71) argues a common core on which all 'richly diverse views of freedom' can agree is that 'freedom is diminished when an agent's purposes are prevented from being realised'. The political and legal structures of liberal societies usually enshrine the principle that individuals are the best judges of their interests. They therefore create a framework in which individuals can pursue their own purposes without undue interference from either the state or from other citizens. In any moderately liberal country, to countenance greater freedom is also to countenance the potential freedom to adversely infringe the freedom of others. This relates to the freedom to commit a crime or to transcend a level which society has deemed formally (by law) or informally (by tradition or custom) to be detrimental to the general well-being of that society. Freedom therefore carries a (moral) responsibility for individuals not to engage in anti-social or criminal behaviour. Indeed, without morally responsible people, it would be very difficult to have a free society. Thus, what must also be considered is the degree to which individuals are held responsible and accountable for their actions and conduct.

As it prevents an agent's purposes from being realised, the fear which stops some social groups from using the city centre is an example of an inhibition on freedom. In the competition for the scarce spatial resources of the city centre, the young and unruly – and predominantly male – are wresting control of the city centre from the rest of the population. Valentine (1990, p.300), for example, notes that men are usually in control of public space at night, both numerically and by their appearance and behaviour. The behaviour of this group frequently infringes the freedom of women. In Britain, night-time activity in the city centre remains dominated by pubs and clubs. The large quantities of alcohol consumed makes people's behaviour even more unpredictable and disturbing to others. As a result those with effective choice are electing not to use the city centre so entraining a spiral of decline which makes the situation worse for all social groups. In a more pernicious spiral, traders and companies compete for shares of a shrinking market, with pubs, clubs and leisure facilities increasingly focused at those groups – and only those groups – who continue to brave the city centre.

Nuisance and Incivilities

The types of crime occurring in city centres typically include shoplifting, pick-pocketing and purse snatching, criminal damage and vandalism, non-residential and some residential burglary, damage to vehicles and vehicle theft, ram-raiding of shops, minor assaults and muggings, affrays, etc. There is also a whole host of offences which may be termed public disorder and nuisance, and social incivilities, which, while not necessarily constituting criminal offences, provoke anxiety and apprehension (**see Table 2.1**). Much of the conduct and behaviour that deters many groups from using city centres is not technically a crime. As Brantingham & Brantingham (1991, p.244) note:

> Civilians see 'crime' as behaviour they find objectionable, regardless of its legal status. Yet the police are held responsible for controlling *legally* objectionable behaviour and may, themselves, be guilty of misconduct if they seek to control legal but otherwise obnoxious forms of behaviour.

The definition of a 'crime' for record purposes by, for example, the police, tends to include only those above a certain threshold of judicial gravity. In their discussion of the 1992 British Crime Survey, Mayhew *et al.* (1993, p.4)

Table 2.1 Common crimes and incivilities in the public space of city centres.

Crime	**Personal/ victim crime** • physical assault • street robbery (mugging and pick-pocketing) • drug-related offences **Property crime** • car theft • damage to car • vandalism • graffiti
Social incivilities (nuisance)	• threatening youths • abusive vagrants/beggars • abusive drunks/ drunks • taunting and harassing groups of children or young people • illegal street vendors • buskers • soliciting prostitutes • drug dealing
Physical incivilities (pollution)	• litter • dogs and people fouling streets • derelict buildings • presence of vagrants/ beggars • presence of winos/ drunks • groups of children or young people hanging around • prostitutes • drug dealers

note that the police use an *operational* definition of crime counting only those incidents which: (i) could be punished by courts; (ii) are felt to merit the attention of the criminal justice system; and (iii) meet organisational demands for reasonable evidence. (See Chapter 6 for further discussion of the police role in creating safer city centres.) Consequently, police statistics do not generally record a number of other public nuisances:

> The term nuisance is widely used as a convenient shorthand for members of the public being upset in any way whilst shopping. This may include shoppers being subject to behaviour which causes them distress, or it may involve the unwelcome and unwanted presence of certain persons in the shopping environment. It may be manifest through unpleasant or offensive gangs of threatening youths; children hanging around in town centres and shopping centres; people selling things on the street; and distress caused by prostitutes, vagrants, beggars, drunks or buskers. (Beck & Willis, 1994, p.31).

Nuisances, such as those described above, are also known as 'incivilities', defined by La Grange *et al.* (1992, p.312) as: 'low level breaches of community standards that signal an erosion of conventionally accepted norms and values'. They also distinguish between *social* incivilities and *physical* incivilities. For the latter, Beck and Willis (1994, p.31) use the term 'pollution' which also includes vagrants and drunks who 'are seen in themselves, irrespective of their actions, as contaminating the environment'. Physical incivilities are discussed later in this chapter. Jane Jacobs (1961, p.39) colourfully refers to what can be considered social incivilities as 'street barbarism' – a phrase which poignantly contrasts with the notion of cities as centres of civilisation – and suggests they indicate that local social control mechanisms are not operating effectively. To address incivilities, she advocates a 'public space civilising service' (Jacobs, 1961, p.51) (see also Gordon & Riger, 1989, p.55; Lewis & Maxfield, 1980, p.160–189).

Social incivilities, jostling, taunting and other harassing behaviour are particularly distressing to women. Junger (1987, p.363) notes:

> Apart from rape, women are exposed to a variety of acts of sexual violence, which vary from relatively harmless, but unpleasant sexual remarks, through all forms of physical (and sexually loaded) contacts and sexual proposals with more or less moral or physical coercion, to very severe forms of sexual violence. The majority of these events which we call 'sexual harassment' do not constitute offences as defined by criminal law and are usually not included in victimisation studies . . . Sexual harassment, in all its forms, could contribute to create a climate in which all women feel less safe than men.

Alongside the more obvious criminal acts, it is these social incivilities which are of concern in this book. For the purposes of this book, while both will usually be referred to as 'crime', it is important to appreciate that the term crime is more widely defined here than commonly used.

THE MOTIVATIONAL DIMENSION

Theories of criminality typically develop accounts of offender motivation and the propensity to crime. Individuals commit crimes for a variety of reasons. There are the highly 'instrumental', acquisitive motivations such as greed or need where the potential gain is usually material and financial. There are also 'affective' motivations where the potential gains are less materialistic and more expressive, such as 'rage' (see, for example, Campbell, 1996); the thrill or excitement of the deed; the wreaking of vengeance or retribution; the show of power or domination over other people; a means of dealing with frustration or gaining sexual release, social approval or friendship. Although it remains debatable, affective motivations are usually less deliberate, less planned, less 'rational', and more emotional and opportunistic.

The Victorian view was that there was a criminal class – the dangerous classes – distinct from a non-criminal class and that crime was directly associated with criminality. Attention, therefore, focused on identifying the different demographical, social or genetic characteristics of each. The modern view generally distinguishes between consideration of the *propensity* to crime (theories of criminality) and the *committal* of the criminal act (theories of crime). It observes that there is a latent criminality in everybody which may or may not result in criminal or illegal activity. In terms of criminal motivation, some people, for reasons such as social position, genetic make-up, or social and economic pressure, will rarely be motivated to crime despite frequent opportunity or temptation. Equally, some people are habitually motivated or pressured to commit crime whenever there is an opportunity. These are however extremes and individuals form some kind of distribution between these poles.

While accepting that there are short term fluctuations in criminal motivation, it is significant to note that criminologists have also identified relatively stable differences among individuals in their propensity to engage in criminal acts (Hirschi, 1986, p.114). Furthermore, there is a certain stability over time of differences across people in their likelihood of arrest (Cornish & Clarke, 1986, p.105). What is also significant is that a small proportion of *offenders* commit a large proportion of *offences*. Bottoms (1990, p.14), for example, cites a study by the Home Office Statistical Department (Home Office, 1985) which found that by the age of twenty-eight nearly one in three males born in 1953 had been convicted of a standard list of offences. Only 0.5%, however, had six or more court appearances before this age, and they accounted for 70% of all known offences committed by the cohort. If these individuals could be identified and removed from society or educated and socialised at an early age, then a significant proportion of crime could be eradicated.

Motivation for Crime

Attitudes towards what is – and what is not – 'acceptable behaviour' depend to a substantial degree upon the assessment of human nature and human motivations. This broadly accounts for the differing attitudes between the political right and the political left. Political attitudes to crime and to criminals are important as they are reflected in the legal and penal response to crime which establishes the judicial and, thereby – in a black letter law interpretation – 'acceptable' parameters of behaviour within a society.

The Political Right

The political right's attitude to crime and criminal behaviour is generally authoritarian and tends to follow a pessimistic outlook on human nature. The argument is that people have deviant tendencies which are kept in check by the presence of self-discipline; various social institutions, for example, the church, family life, the law, and those entrusted with authority. Their view therefore tends to focus on internal motivations for crime. Thus, crime 'occurs because of a lack of conditioning into values: the criminal . . . lacks the virtues which keep us all honest and upright' (Lea & Young, 1984, p.136).

Concern is also placed most forcefully upon the victim of the crime. In the UK, Conservative Party policy, for example, 'relates more to the victim rather than the perpetrator, who is viewed as a maladjusted delinquent in need of severe punishment in order to direct him along the correct path' (Jones, 1988). Thus, it is argued, where self-discipline or self-control is lacking, harsh sanctions are required to deter that individual. The political right has therefore tended to argue for longer prison sentences and for capital and corporal punishment. In the US, California has a 'three strikes' policy whereby an offender convicted for the third time receives a mandatory life sentence.

An alternative right view – the 'radical right' – sees crime as stemming less from a lack of training in values and morality, and more from innate evil and psychotic maladjustment. McLaughlin & Muncie (1996, p.2), for example, note the emergence in the late 1970s of an authoritarian populist New Right discourse which 'rested on cherished memories of a safer, more orderly society and a hard-hitting analysis on how to wrest society back from the criminals, terrorists, strikers, muggers and hooligans'. The New Right pinned the blame firmly on:

> irresponsible policies which had (a) not only denied but sought to erase all notions of human wickedness, right and wrong, self-discipline, personal responsibility and shame; and (b) undermined the core institutional arrangements and cultural precepts that were the essential sources of social order, discipline and shared moral meaning. (McLaughlin & Muncie, 1996, p.2).

The Political Left

By contrast, the more idealist on the political left generally hold a more optimistic and tolerant view of human nature. Their view tends to focus on external motivations for crime: 'crime occurs not because of lack of values but simply because [of the] lack of material goods: economic deprivation drives people into crime' (Lea & Young, 1984, p.136). The motivation towards crime is also attributed to the general inequality and powerlessness of many in society and the frustration, resentment and envy generated, particularly if expectations cannot be met legitimately.

Some critics argue that the left idealist viewpoint legitimises criminal behaviour as the only means of existence for individuals, who – if given the chance in life enjoyed by the middle and upper classes – would be law-abiding citizens (Crowe, 1991, p.8). Critics also suggest that the viewpoint is a form of economic determinism, neglects the role of human agency and moral responsibility, and therefore fails to levy appropriate blame on the perpetrator.

An alternative left view – *left realism* – argues that the Left has 'traditionally either romanticised or underestimated the nature and impact of crime and largely "speaks to itself" through its lack of engagement with the day-to-day issues of crime control and social policy' (Muncie *et al.* 1996, p.xxi). In seeing people's fear of crime as rational and a reflection of inner-city social reality (Jones *et al.*, 1986), left realism mirrors right realism. Nevertheless, as Muncie *et al.* (1996, p.xxi) argue, it differs in its insistence that 'the causes of crime need to be once more established and theorized; and a social justice and welfare programme initiated to tackle social and economic inequalities, under the rubric of "inclusive citizenship"'.

Depending on the particular political background, the increase in crime and anti-social conduct in the post-war period is attributed variously to the permissive society of the 1960s and/or the economic depressions, unemployment and polarisation of society since the 1970s. More likely it is a combination of each. In general terms, it is possible to argue that with increased unemployment and poverty (which has tended to increase since the early 1970s) more acquisitive and instrumental 'crime' in the form of burglary and robbery may be expected. The opportunities for such crime have also increased following the economic expansion and the increasing demand for consumer goods in the post-war period. Equally, the breakdown of family life and of a generally accepted social order and responsibility (for example, since the 1960s) may be expected to lead to increased affective crime, social disorder and nuisance. Research in this area is as yet inconclusive and clouded by issues of political partiality and ideology.

Other factors will also contribute towards the motivation to indulge in crime. Prevailing cultural and sub-cultural factors, for example, may help explain greater incidences of crime in some societies. In the UK, for example, both Hope (1985) and the COMEDIA *Out of Hours* report (1991), for example, suggest that the root causes of a culture of heavy drinking and violence need to be addressed. Hope noted that: 'there are strong associations between youth, drink, disorderly conduct and gender, as well as situational factors of space, public domain and physical structure'. To this may be added the role increasingly played by drugs. Harrison (1983, p.335), for example, discusses the pernicious attractions to teenagers of a 'sub-culture of conspicuous consumption'; participation in which requires money often only available through crime. Similar peer pressure is seen in many US cities in, for example, the wearing of the latest – and often very expensive – designer sports shoes.

THE VICTIM DIMENSION

The Incidence of Crime

Estimates of the number of offences recorded by the 1992 British Crime Survey (BCS) are shown in Table 2.2. The figures for violent crime are perhaps most appropriate in considering the impact of crime in city centres. The most numerous type of assaults uncovered by the BCS were incidents of 'domestic violence', eight out of ten of which were against women; emphasising the fact that for many women their own home is not a safe haven. Mayhew *et al.* (1993, p.83) estimate that the survey figures suggest a minimum of 530,000 incidents of this nature in England and Wales in 1991. Street assaults are nearly as common as domestic assaults, amounting to about 510,000 incidents in 1991; eight out of ten of which involved men. There were an estimated 420,000 assaults in pubs and clubs – eight out of ten of which were against men – and some 260,000 incidents of 'mugging'. Muggings were evenly split between male and female victims, though women were more often subject to snatch theft and men to robbery (Mayhew *et al.*, 1993, p.83).

In terms of the timing of violent crimes, the survey found that violent incidents were more likely to occur at the weekend, with half of all incidents

Table 2.2 British Crime Survey estimates of crimes against individuals and properties (source: Mayhew *et al.*, 1993).

TYPE OF CRIME	Number of offences	%
Home vandalism	1,063,000	7.1
Burglary	1,365,000	9.1
Theft from person	439,000	2.9
Common assault	1,757,000	11.7
Violent crime	809,000	5.4
Other thefts	4,146,000	27.5
Vehicle crime:	5,476,000	36.4
Vandalism	1,668,900	30.5
Theft of vehicles	517,400	9.4
Theft from vehicles	2,339,600	43.8
Attempted theft from vehicles	889,800	16.2

taking place between 6pm on Friday and 6am on Monday. Not surprisingly, pub violence most often took place on Friday and Saturday nights, but more than half of the domestic violence also happened during the weekend. Within this overall time frame, most violence occurred in the evening between 6pm and midnight, particularly pub violence (80%). Women were more likely, however, to be mugged during the day than men (51% as against 18%) – perhaps because fewer are out after dark. Although the question was not asked in the 1992 survey, the contribution of alcohol had been firmly established in the 1988 survey where: 'In four out of ten incidents of violence . . . the victim said the attacker was drunk' (Mayhew *et al.*, 1993, p.92).

Regarding the impact on victims of crime, apart from physical injuries, many victims experienced emotional reactions ranging from shock, anger and fear to difficulty with sleeping and depression. The most common reaction for both male (44%) and female (58%) victims was anger, followed by shock (25% of male victims, 47% of female). Women – particularly elderly women (although the numbers were small) – were more likely to report such reactions than men. Women were most upset by domestic assaults and muggings, though men were even more affected by muggings than women (more than half reporting they were 'very much affected'). Victims were generally more upset by attacks from those they knew than by strangers (Mayhew *et al.*, 1993, p.94).

The Risk of Victimisation

What is arguably of more concern to people than the rates of crime is the incidence of crime in areas they are likely to use and the risk of crime to them

Table 2.3 Relative factors in the risk of violence from the British Crime Survey. Rates are indexed at the pooled rate for the full 1988 and 1992 BCS samples, except those marked * which were not asked of the full 1988 and 1992 sample (source: Mayhew *et al.*, 1993).

RELATIVE FACTORS IN THE RISK OF VIOLENCE		
ACORN risk group	Low Medium High	75 130 160
Number of evenings out in last week *	None or one Two or three Four or more	65 125 200
Visited pub in last week *	Yes No	195 95
Children under 16 in household	Yes No	130 85
Age	16 - 29 year olds 30 - 59 year olds 60+ year olds	280 60 10
Male		130
Female		75
Men, aged	16 - 29 30 - 59 60+	360 70 15
Women, aged	16 - 29 30 - 59 60+	200 50 10
Inner city		130
Non inner city		95
Number of adults in household	One Two Three or more	125 65 150
Drinking behaviour	None A little Moderate/heavy	75 70 140
Incivilities scale	Low High	85 335
Household income	Higher Medium Lower	70 105 115
Married		50
Single		270
Separated/divorced		210
Widowed		15
White		95
Black		160
Indian/Pakistani/Bangladeshi		100

as individuals and families. The 1992 British Crime Survey also assessed the risk of violent crime for a range of variables relative to the national average incidence risk (see **Table 2.3**). As would be expected, people's age, gender, home location, ethnicity, wealth, and lifestyle choices affected their risk of being a victim of violent crime. Age is one of the most significant factors in an individual's risk of being a victim of crime (Mayhew *et al.*, 1993, p.80). The older people were, the lower their risk. The risk of being a victim of violent crime for those aged 16 to 29 is about two and a half times the national average; while the risk for those people of 60 plus is a tenth of the national average. The latter may also be explained by the fact that the elderly often take precautions, thereby putting themselves less at risk.

Fear of Crime and Victimisation

In terms of the impact of crime on urban life and urban living, the perception of crime is often more important than the actuality. As many people, especially women, limit their activities because of fear of crime, fear is as great a problem as crime itself. The 1989 Home Office report, *The Fear of Crime*, for example, emphasises that fear of crime can be as instrumental as actual victimisation in lowering the quality of people's lives. Perception – rather than reality – is the conditioning factor in everyday experience and therefore behaviour. A crowd of youths may simply be noisy and good humoured, but desired behaviour may have to be modified due to the on-looker's perception of the possibility of deleterious conduct.

Fear of crime also may be out of all proportion with the risk of victimisation. There is, however, a convincing explanation: as vulnerable people will tend to be more cautious and risk averse, they will take precautionary measures, including placing curfews on themselves. Thus, as a direct result of fewer women and older people putting themselves at risk, they are less well-represented in the number of victims of crimes. As a consequence, crime statistics do not show them to be most at risk. Indeed, as Gottfredson (1984, p.31) observes, according to the crime statistics, the high risk individuals are 'young, urban, working class males who are out in public or inside public houses a great deal'.

Lotz (1979, p.241–54) distinguishes between 'fear' of crime and 'concern' about crime. To her, fear denotes an individual's perceived chances of being a victim, while concern stands for one's estimate of the relative seriousness of crime nationally. Often there is a significant gap between people's fear of crime and the actual incidence of crime. For example, the 1981 British Crime Survey stated that a statistically average person of sixteen years or older can expect to experience robbery (not attempts) once every five years; an assault resulting even in slight injury once every century; the family car to be stolen every sixty years; and a burglary in the house once every forty years (from Jones, 1989, p.304). On these figures, most people can expect to experience relatively little crime during their life time. Nevertheless, the *Islington Crime Survey* found that many crimes are not captured by national surveys and that, in some areas, the fearful have every reason to be so (see Jones *et al.*, 1986).

Table 2.4 Fear of crime: summary of surveys (from Atkins, 1989, p.8)

FEAR OF CRIME : SUMMARY OF SURVEYS		
Location	Date	Finding
London	1984-85	56 per cent of women felt 'very unsafe' or 'not very safe' walking alone at night; 22 per cent never travelled after dark
Islington	1985	73 per cent of women and 27 per cent of men felt worried about going out alone at night
Manchester	1987	63 per cent of women never walked home alone at night
Birmingham	1987	69 per cent (both sexes) deterred from visiting city centre at night
Southampton	1986	59 per cent of women felt unsafe walking after dark; over 90 locations identified as unsafe by respondents
Lewisham	1985	53 per cent of respondents felt unsafe going out at night on one estate; 79 per cent on another
Croydon	1986	'Almost two-thirds' of respondents did not feel safe walking alone after dark; 'over half' in another location
Swansea	1986	Half the respondents felt unsafe when walking about estate after dark
Wellingborough	1986	58 per cent felt they would be victims of violent street robbery; 54 per cent felt they would be assaulted in the street; 60 per cent of women thought they would be sexually assaulted
Great Britain	1987	40 per cent of respondents feared going out at night; 59 per cent of retired persons; 64 per cent of women
England and Wales	1984	31 per cent of respondents felt 'fairly unsafe' or 'very unsafe' walking alone after dark

A number of surveys clearly indicate that women and, to a lesser extent, older men, are apprehensive about living in and/or using city centres (**Table 2.4**). A study in the USA (CCC, 1985, p.5), showed that people are most afraid of being mugged, raped or assaulted, and of having their cars broken into or stolen. Sixty-three per cent of the respondents said they rarely used the city centre or walked in the streets because of fear. Similar concerns have also been reported in the UK. A survey carried out for the Nottingham Safer Cities Project (1990) interviewed nearly a thousand people aged over sixteen about their fear of crime, perception of risk and avoidance behaviour. The survey revealed that a substantial proportion of respondents avoided the city centre after dark. Furthermore, 45% of those who did use it in the evenings, said they felt unsafe. Some – 3% on weekdays and 7% at weekends – even avoided the centre in the day-time through fear of crime. About half of those questioned said they were worried about mugging and being attacked by strangers. The most common reasons given for feeling unsafe were 'people hanging about' (57%) and a general fear that 'something may happen' (38%). The respondents mentioned 27 specific places that they would avoid, including the Old Market Square – Nottingham's main public space.

Guessoum-Benderbouz's survey of women in Nottingham in 1994 replicates and reinforces many of the findings of the earlier survey. Based on over 400 postal returns, the results showed that 63% of women over eighteen felt unsafe in the city centre. Nearly a quarter of the respondents had not used the city centre in the previous month, despite it being a vital part of their lives: 87% used it for shopping and 16% worked in the city centre. Although the majority (over 80%) do not refrain from using the city centre during the day, only 14% felt safe after dark. Thirty per cent of women do not go out after dark, 48% avoid the city centre after dark and a further 8% avoid the city centre in the late afternoon or early morning. As one in seven of the respondents to the survey had been a victim of any crime or assault in the city centre, the scale of the women's fear may be seen as disproportionate (although one in seven is still a very high proportion). Those who had been victims tended to come from lower socio-economic groups, poorer areas of the city and tended to be younger than the average for the survey. It is these younger women who appear to be reluctant to put curfews on themselves and seem to suffer the consequences. Of the sixty crimes reported by respondents, only twenty-five had occurred in the city centre. Nevertheless, the impact of these experiences on the women concerned was significant: two of the victims gave up work, four gave up social outings altogether, fifteen will go out only in the company of others and eight changed their routines. The impact of such incidents results in large numbers of other women curtailing their activities or, even if they use the city centre, being uncomfortable there. Even the 14% of the respondents who stated that they feel safe and comfortable in the city centre after dark, do not use the city centre alone after dark. What they may actually mean is that they feel comfortable when accompanied by others.

Factors in the Fear of Crime

Skogan (1984, from Junger, 1987, p.360) argues that the fear of crime has three components: cognitive, evaluative and emotional. As the cognitive and emotional dimensions are usually most significant, the objective assessment of actual risk is distorted and often exaggerated. This is not, however, to pathologise fear of crime as 'neurotic' or 'irrational'. A number of factors contribute to the fear of crime, such as vulnerability, perceptions of personal risk and the seriousness of various offences; environmental clues and conditions; personal knowledge of crime and victimisation; confidence in the police and criminal justice systems (Box *et al.* 1988, p.341).

Vulnerability and Perceptions of Personal Risk

Fear of crime is more pronounced for those who feel unable to protect themselves physically or economically, or who feel less able to cope with the physical and emotional consequences of being a victim (Toseland, 1982). One factor is an individual's own physical strength, power and ability compared with that of possible attackers. The judgement as to what is and is not safe depends on this comparison. As the average woman is weaker than the average man, they have less confidence in their ability to defend themselves and are less likely to put themselves into a situation where it might become necessary. Maxfield (1987) emphasises the distinction between *feeling* vulnerable and actually *being* vulnerable, while Box *et al.*, (1988, p.341) note that research has identified four groups that feel particularly vulnerable: the old, women, the poor, and ethnic minorities.

Women of all ages are more vulnerable to the effects of crime due to their additional vulnerability to rape and sexual violence and harassment (see for example Stanko, 1985). Although the boundaries between harassment and violence can be hazy, sexual harassment is usually defined as largely non-physical threatening behaviour, while violence is held to involve physical violence and abuse (Pain, 1995, p.589). Pain found that elderly women tend to worry more than younger women about sexual harassment. In her survey, around half the women in all age groups worry about receiving obscene phone calls, younger women were the most likely to worry about being followed, and that women over 60 were more likely than all other age groups to report being 'fairly worried' or 'very worried' about being flashed, leered, or whistled at (Pain, 1995, p.588–589).

In general, women are more fearful of victimisation than men, but, as men grow older, the gender-fear gap narrows. Men do not, however, ever become as fearful as women. Box *et al.* (1988, p.352) suggest that:

> This levelling process may be because age gnaws the strongest spirit, even among men once proud of their physique and strength. In addition, to this increasing vulnerability, men, compared to women, may have practised fewer of the subtle arts of offender avoidance.

Environmental Cues and Conditions

Fears of a lack of safety in certain spaces are also given by environmental signals or clues. Box *et al.* (1988, p.342) note that physical incivilities, neighbourhood housing conditions, and the lack of a sense of neighbourhood cohesion and community all contribute to the fear of crime. Similarly Lewis & Maxfield (1980) found that signs of social disorganisation affect people's perceptions of their neighbourhood's crime problem and consequently their reactions to crime. As Kinsey *et al.* (1986) and others note, the more vulnerable citizens of inner cities – women and older people – perceive crime as an area problem and fear certain environments. Many people are apprehensive about or will fear at least certain parts of the city centre, such as pedestrian subways or dark alleys. Many are also disturbed by situations which restrict choice or offer no alternatives: for example, subways as the only means of crossing busy roads; narrow pavements and constricted entrances, particularly when these are already crowded by – what the Americans term – 'people who create anxiety'. Such incivilities evoke fear in people and encourage them to take precautions.

Table 2.5 Design characteristics that contribute to women feeling 'unsafe/ uncomfortable' (source: Whitzman, 1992 from METRAC).

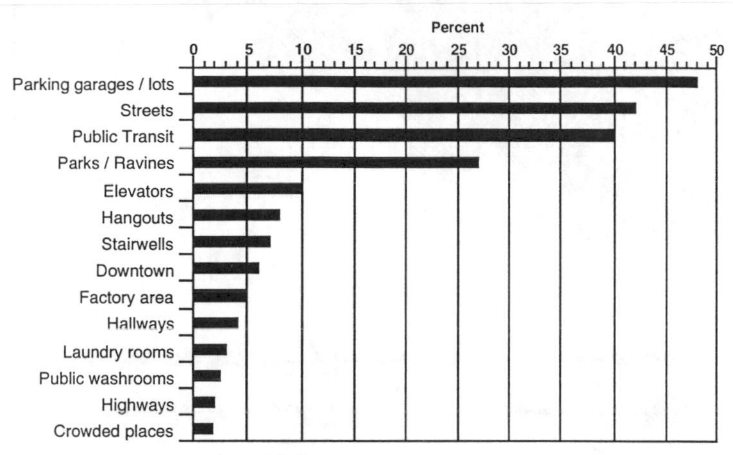

Women are more sensitive than men to cues of social and physical disorder. Montgomery (1994, p.304), for example, argues that women in particular construct mental maps of parts of cities distinguishing between those that feel safe and those that do not and should be avoided. Wekerle & Whitzman (1995, p.4) note how William H. Whyte, in his studies of plazas and street life, concluded that: 'women's desertion of certain public places in cities is a signal, just like the canaries in mines, that the place is in trouble'. They also observe that most women will spontaneously identify what criminologists call 'hot spots' of predatory crime and fear (**see Tables 2.5 & 2.6**):

> When asked to identify 'dangerous places' in the city or their own neighbourhood, over three-quarters of all women name very specific places where they take special precautions. They identify certain streets or alleys, parks, deserted places, public transit, parking garages, and elevators. When asked what it is about the places that makes them feel unsafe women are very specific: poor lighting, places that are isolated or deserted, and places where there is no access to other people. (Wekerle & Whitzman 1995, p.4)

Table 2.6 Public places where women feel 'unsafe/uncomfortable after dark' (source: Whitzman, from METRAC).

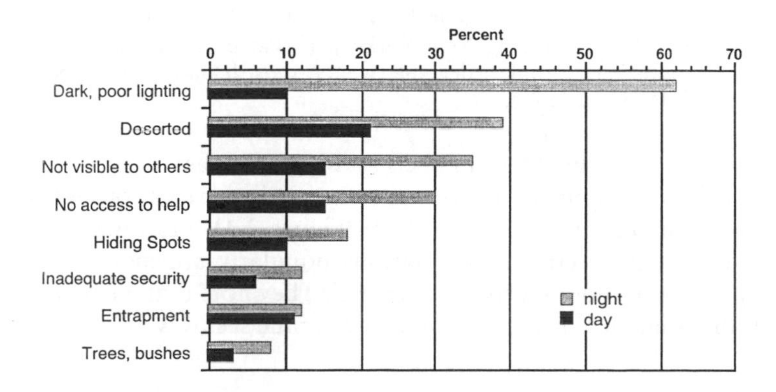

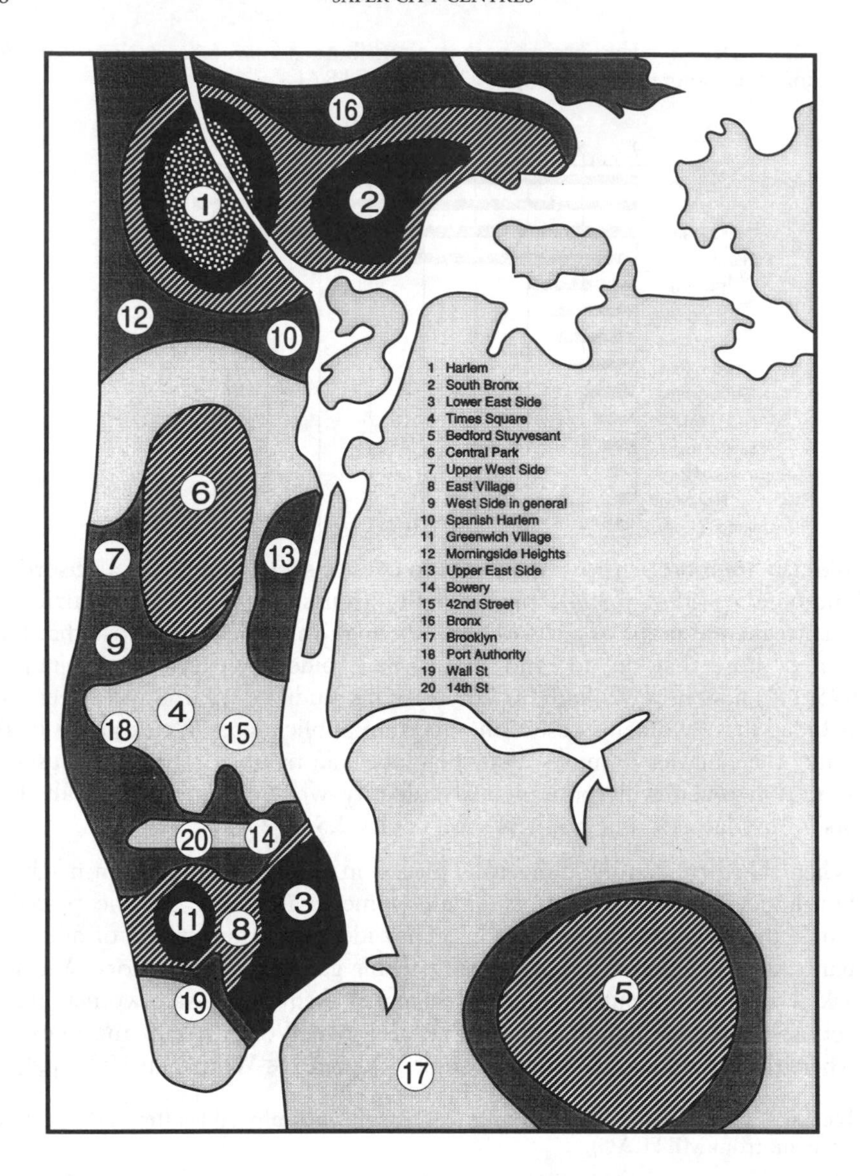

Figure 2.1 The topography of fear in New York 1977. The darker the shading, the more widespread the fear among New Yorkers of that particular part of town. The numbered locations list, in rank order, the twenty 'most fearsome' neighbourhoods in the city (source: Duncan, 1977).

Ecological Labels and Reputations

What will also contribute to fears of crime in particular locations is the reputation or popular image attached to them (**Figure 2.1**). Brantingham & Brantingham (1991, p.4) term the reputations popularly appended to particular places or neighbourhoods 'ecological labels'. They note that what is significant is that rather than actual crime rates, fear of crime seems wedded to ecological labels. What is also difficult is that negative perceptions of areas can be particularly enduring.

One of the major factors in establishing the reputations of areas and locations is the media and other received information. Although, information on crimes and other nuisances can come from a variety of sources – not least from neighbours, friends and relatives with whatever degree of exaggeration and hyperbole[1] – it is the media which play the most significant role in exacerbating people's concerns about crime levels and their fear of crime. The media are often portrayed – in Stanley Baldwin's phrase borrowed from his cousin, Rudyard Kipling – as possessing 'power without responsibility'. But, whether responsible or irresponsible, all newspapers are inevitably selective and tend to favour the sensational and extraordinary over the mundane and everyday. Research (see, for example, Ditton & Duffy, 1983; Marsh, 1991; Williams & Dickinson, 1993; Heath, 1994) has found that the newspapers tend to have an over-emphasis on crimes of violence and crimes involving indecency. Box *et al.* (1988, p.342) note that when the press portrays an image of crime as irrational 'random choice of victims, normlessness accompanying criminal behaviour, and dramatisation of events and victimisation risks – it contributes to fear amongst its readers'. Thus, the reporting – and sometimes misreporting – of crime in the local and popular press can often be responsible for the persistence of mis- or half-truths in the popular imagination.

The media may also help fuel what are termed 'moral panics'. As Muncie (1996, p.50) notes, the concept is used to describe public reactions – or, perhaps more pertinently, media and political reactions – to various anti-social or deviant activities. He identifies three periods of moral panics: the 'discrete moral panics' of the period 1955–65 (e.g., teddy boys; mods and rockers); the 'more diffuse moral panics' of the period 1966–70 (e.g., permissiveness; drugs; student radicalism) and the 'generalised climate of hostility to marginal groups and racial minorities' since 1971 (e.g., street crime; drugs; family morality; sub-cultures, such as punk, acid house and rave parties, the 'yob' culture). The response to moral panics can be interpreted in different ways. Some see the response to moral panics as relatively benign, being simply the need for society to make periodically new 'rules'. By contrast, those who regard crime as resulting from the state's power – or desire – to criminalise, see moral panics as more pathological and interpret them as an attempt to demonise certain groups or to portray them as contemporary 'folk devils'. To them a moral panic represents an over-reaction to the actual threat posed and/or an over-exaggerated presentation of the impression of a crime wave, which thereby justifies more punitive sanctions. Thus, the moral panic, as Muncie (1996, p.53) observes, is 'the first link in a spiral of events leading to the maintenance of order in a society by a legitimised rule through coercion and the general exercise of authority'. Some commentators, for example Young (1974), also argue that the media's need for 'good copy' encourages them to engineer periodic moral panics.

1 Agnew (1995, cited in Box, *et al.*, 1988, p.342) notes that victims may neutralise the impact of crime so that they are able to preserve their emotional balance and continue functioning in a relatively normal way.

Confidence in the Police and Other Public Guardians

In general, if areas are well managed, cared for and controlled they are perceived to be safer. This sense of security is reinforced if there is a visible human presence. Respondents to Guessoum-Benderbouz's survey of Nottingham, for example, wanted a greater police presence and protection, while as Box *et al.* (1988, p.342) note:

> If people believe that the police are effective and efficient at clearing-up crimes and apprehending criminals, that they respond to calls quickly and that they have physical presence on the ground, then they are less likely to fear crime . . . Thus, confidence in the police becomes another factor facilitating or muting the development of fear.

The human presence may also be those with a managerial responsibility such as uniformed station attendants or bus conductors (see for example Sturman, in Clarke & Mayhew (1980), on the presence of bus conductors in inhibiting vandalism). Increasingly on private property, there are also security guards.

Precautionary Strategies

Perceived fear imposes a significant psychological cost and seriously inhibits the behaviour of a large number of people. Fear of crime is a rational response to the apparent irrationality of crime and criminal behaviour. Jane Jacobs (1961, p.40) colourfully observed:

> To be sure there are people with hobgoblins in their heads, and such people will never feel safe no matter what the objective circumstances are. But this is a different matter from the fear that besets normally prudent, tolerant, and cheerful people who show nothing more than common sense in refusing to venture after dark – or in a few places, by day – into streets where they may well be assaulted, unseen or unrescued until too late.

Those who feel vulnerable to crime are likely to be risk-averse and will take precautions in order to lessen the risk of their falling victim to criminal behaviour. Thus, Garofolo (1982, p.856) argues given current realities 'fear is functional to the extent that it leads people to take reasonable precautions'.

In general, there are two main precautionary strategies: 'avoiding dangerous situations (e.g., not going out alone at night) and managing risks in the face of possible danger (e.g., by asking repairmen for identification)' (Riger *et al.*, 1982, p.369). Each strategy may be exercised by the same person at different times. A third strategy may be to practise denial.

Risk Avoidance

Avoiding areas identified as potentially dangerous is a rational human reaction and, given choices, many people choose for example not to use city centre streets after dark. Rather than *actual* fear, it is a response to *anticipated* fear.

Avoidance refers to those actions taken to decrease the exposure to crime by removing or distancing oneself from situations in which the risk of victimisation is perceived to be high (DuBow *et al.*, 1979, p.31). The dimensions of situations to be avoided are various: location, time of day, people, etc. As fear for personal safety is a major factor in limiting personal mobility, avoidance can lead to a 'fortress mentality' which impoverishes the person's quality of life. Avoidance also has an economic and cultural cost as those practising avoidance remove themselves as both consumers and as citizens.

Risk Management

There are many groups for whom risk avoidance is not possible or desirable. Many people lack the financial means to opt out or buy avoidance; for example, those who do not own or have the use of a car are forced to rely on public transport. Similarly, structural constraints and the role obligations dictated by lifestyles and necessary routine daily activities circumscribe people's ability to use precautionary tactics and increase the probability of their being in places at times when victimisations are known to occur (Riger *et al.*, 1982, p.373). Thus, for example, there are significant numbers of people, many of them women, who have to be in city centres after dark or before 9 am as a condition of their employment, notably the large army of female cleaners.

People in this group have to practise risk management: 'what people do to deal with the perceived risks when they cannot or will not physically avoid them' (DuBow *et al.*, 1979, p.41). Risk management is for the most part a response to anticipated fear, but may – on occasion – also be a response to actual fear. Risk management involves a number of different strategies: risk reduction and/or spreading, risk transfer, and risk retention, and seeks to minimise or manage the risk of victimisation in the face of danger, by, for example, learning self-defence, carrying a weapon, or not making eye contact with threatening strangers. As Riger *et al.* (1982, p.371) note risk management strategies: 'range from active assertive behaviours, such as carrying a gun for self-defense, to more passive restrictive behaviours, such as never going out alone at night'.

In Britain, the government has also sought to encourage people to take more responsibility for their own crime protection. The Home Office (1991) publication, *Practical Ways to Crack Crime*, for example, gave precautions on how to help people, and women in particular, feel safer and more secure in their everyday life. As Stanko (1990) has argued the irony in this advice is that most women do not need to be told about taking such precautions – they have already devised many of their own. Without denying the 'good sense' and the importance of precautions, the advice tends to mask the fact that there is a wealth of evidence from feminist and other research that women are in greatest danger from men they know (Walklate, 1996, p.300). The Home Office (1991, p.5) advice also suggested what men can do to help by taking care not to frighten women. These include not walking behind women on their own; not sitting too close to a woman on her own in a railway carriage; and remembering that a woman on her own may feel threatened by what a man thinks are admiring looks.

Risk management strategies have similarities with those of the situational approach to crime prevention outlined in the next chapter. Risk management may also involve 'insurance behaviour' – behaviour which seeks to minimise the costs of victimisation or alters the consequences of victimisation. It is extremely doubtful, however, whether subsequent financial recompense can ever wholly make redress for the psychological consequences of victimisation.

In any civilised society, there are limits to which personal freedoms and behaviour can – or should – be restricted in order to resist crime and restrict anti-social behaviour. Other options may remain outside the scope of personal action, necessitating, for example, modifications to the created environment (the 'opportunity') and changes in other people's conduct (their 'motivation'). Furthermore, precautionary strategies are 'private-minded behaviours'; they protect only the individual who adopts them, not everyone in the immediate neighbourhood ('public-minded behaviour') (Riger *et al.*, 1982, p.371). Their successful implementation may lead to the deflection of crime onto other people or places such as those less able to practise or purchase avoidance and other forms of protection. Thus, the individualistic, welfare-maximising behaviour of some people may – unintentionally – make the situation worse for other people (especially the poor who have few options) and even for society as a whole.

The Consequences of Fear of Crime and Victimisation

Crime and the fear of crime manifest themselves in a variety of mostly deleterious ways, but the overarching effect is a reduction in the quality of life for the majority of a city's inhabitants. Box *et al.* (1988, p.340) note that the fear of crime:

> fractures the sense of community and neighbourhood, and transforms some public places into no-go areas; because fear leads to more prosperous citizens protecting themselves and their property, or moving from the neighbourhood, the incidence of crime may be displaced on those already suffering from other social and economic disadvantages; it reduces the appeal of liberal penal policies, such as decarceration and rehabilitation, thus paving the way for more incarceration and punishment; it creates a seedbed of discontent from which vigilante justice might flourish and thus undermines the legitimacy of the criminal justice system, particularly when courts are seen as being soft, displaying more compassion for the offender than the victim.

Crime reduces the profitability of local shops and businesses, increasing their insurance premiums and often making them uninsurable (Harrison, 1983, p.345). It also damages a city centre's economy by inhibiting the behaviour of its users. In the US, the Citizens' Crime Commission (1985, p13) noted that:

> it depresses the multiplier effect by reducing the level of pedestrian activity and the distances people are willing to walk on the streets; it encourages

insulated activity in which self-contained complexes and indoor walkways are preferred to outside sidewalks; it decreases the level of face-to-face communication between users; it promotes the desertion of the area after 5 pm; it increases automobile use and demand for close-by parking.

If city centre streets are perceived to be dangerous, they may be either deserted or abandoned and perhaps given over to gangs of revellers and drunks after dark. If avoidance behaviour continues, further perpetuating the perception that the place is unsafe, a pernicious spiral of decline can set in. Whether as a cause or consequence, this not only denies large numbers of men and even greater numbers of women the use of their city centres at night, but also has a significant economic and employment cost. For Nottingham, it was calculated that there was over £24 million in lost turnover and 462 lost job opportunities (KPMG Peat Marwick, 1990).

More widely, society as a whole is continually paying for the existence of a criminal element. There are the direct costs in the form of the police, prisons and social services which are borne by all tax payers. There are also additional hidden costs, such as the expense of maintaining people who commit offences which is effectively redistributed to all members of society. Hence, the victims of crime are not just the individuals who suffer, but society as a whole.

THE OPPORTUNITY DIMENSION

In analysing the locational dimension of crime the spatial level of analysis is important. Appearances of homogeneity at higher levels may mask important differences in the distribution of crimes, offenders, and targets at lower levels. The analysis can be grouped into three levels: *macro*, *meso* and *micro* (Brantingham & Brantingham, 1991, p.21–22). The macro level refers to studies at the highest level of aggregation, such as the distribution of crime between countries or cities. Research at this level rarely produces findings with preventative implications. The meso level refers to studies at intermediate levels of spatial aggregation, such as the study of crime within sub-areas of cities. The lowest level of analysis, the micro level, refers to studies of specific crime sites and is of most relevance to this book. It is however important to discuss briefly the meso-level.

The Meso Level of Crime Incidence

The geographic incidence of crime is generally uneven; more crime occurs in some areas than in others. The development of urban criminology and the geographical mapping of crime was pioneered by Guerry (1833) and Quetelet (1842), who analysed early French crime statistics and mapped convictions for violent and property crimes at the departmental level. They both found that crime was unevenly distributed across the departments, and that the patterns for violent and property crime were different. In London, Henry Mayhew *et al.* (1862) also mapped the incidence of crime. They recorded the fact of persisting criminal areas – thieves' quarters popularly known as rookeries. Rookeries were like rabbit warrens with numerous escape routes for criminals. They were also positioned to take advantage of the distribution of targets and of important differences in the quality of policing in different precincts. Early slum clearance programmes were, in part, an attempt to remove these.

In the first part of the twentieth century, particularly in the US, there were further studies of spatial criminology. Whereas the nineteenth century research had largely been descriptive, the early twentieth century researchers drew reference from the social ecology theory of the Chicago School. In the 1920s and 1930s, research into the rates of delinquency, carried out in Chicago by Shaw & McKay (1929; 1931; 1942), offered a more systematic approach to the concept of identifying crime and general zonal patterns of crime. Their research was linked to the Burgess model since as early as 1925, Robert Park (1925) and his students had noted that in the transitional zone around the core, crime rates were higher than in the rest of the city. The explanation offered by the Chicago School was that of 'social disorganisation' – the breakdown of rules governing social life:

> Essentially, their argument was that offending manifested itself in a lack of structurally located social bonds which encouraged legitimate and discouraged deviant behaviour. Such social disorganisation was the result of new immigrant populations coming together and not having had the opportunity to develop a stable social structure with clear norms. (Bottoms & Wiles, 1992, p.100).

As applied in criminology, social ecology had two distinct components. The first was a theory of urban form based on the idea of competition or conflict between social groups for territory. This component remains a useful explanation of change within cities. The second component was a social psychology that predicted the nature and quality of social organisation within different natural areas, their impact on individuals and groups, and which has been subject to significant criticism and revision. As Knox (1986, p.61) notes:

> Despite its far-reaching effects on the orientation of a great deal of subsequent work in urban sociology and urban geography, traditional human ecology has been abandoned in favour of a much-modified ecological approach based on the idea of identifying key social variables and examining their relationships within an 'ecosystem' or 'ecological complex'.

More recent studies of the pattern of crime in cities have focused less on social ecology and more on mobility (for example, see Felson, 1986; 1987; Brantingham & Brantingham, 1991; 1993). Brantingham & Brantingham (1991, p.240) argue that crime is highly patterned by daily behaviour and most criminal events can be understood in the context of people's normal movement through normal settings in the course of everyday life. Through routine daily activities, all people – including those who commit crimes – develop an 'awareness space' that builds upon the activity space and its associated spaces. Criminal targets are usually selected from within this awareness space. Thus, offences are committed near the places where offenders spend most of their time – home, work, school, shopping, entertainment – and along the major pathways between them.

The meso level of analysis has a general value but the unit across which statistics are aggregated frequently relates poorly to the pattern of crimes when plotted individually (Barr & Pease, 1992, p.196). The general problem is that the pattern of association between crime and potential causes observed at the aggregate level may not apply to individuals: this is the basis of the 'ecological fallacy'. Conclusions about individuals may then be wrong or at best simply irrelevant (see Bottomley & Pease (1986) for a full discussion on the shortcomings of aggregated crime figures and their relationship with other variables). In the context of this book, what is perhaps more significant than the area's general characteristics is the way in which the conditions of the immediate micro-environment influence the behaviour of individuals.

The Micro Level of Crime

Any consideration of the micro level of crime incidence must consider the influence of the immediate environment on people's behaviour. Environmental determinism is based on the premise that the built environment directly shapes the behaviour of people using it. Such a concept might have had some currency in the extreme conditions of the nineteenth century slums, but the idea remained a potent one throughout the early twentieth century among architects, planners and other built environment professionals. By the mid-1960s, the professionals' tacit acceptance of physical determinism was disputed and contested by arguments that placed their emphasis on cultural, social and economic factors. Nevertheless, to deny that the environment has no influence would be naive. A more realistic approach is to accept that physical factors are not the exclusive influence on behaviour.

Bell *et al.* (1990) point out other perspectives on behaviour in the built environment, such as environmental *possibilism*, which views the environment as presenting individuals with opportunities as well as limiting their behaviour. Behaviour is therefore 'situational' or contextual; within particular physical settings people make – generally rational (albeit a bounded rationality) – choices among the opportunities available to them. Between determinism and possibilism lies environmental *probabilism*. Probabilism assumes that individuals choose a variety of responses in any environmental situation but 'some behaviours are more likely to occur than others' (Bell *et al.*, 1990). Thus, in a given physical setting some choices are more probable than others. The choices are made with regard to individual motivations, values, and cultural norms.

As people generally have some autonomy and can usually modify the immediate conditions of their environment, most needs apparently dependent on the environment may be satisfied in other ways: for example, by the altering economic, social or cultural conditions, or the prevailing managerial regime, or by a combination of these. Where people lack autonomy or ability (for financial, personal or for whatever reasons), the inhibitive or disabling effects of their immediate environment are a principal or dominant influence – in an obstructive manner; they lack the means to transcend its limitations. As needs can be met in *diverse ways*, there is no deterministic relationship between human requirements and physical settings. There are, nevertheless, pragmatic limits to the physical scope of this adjustment beyond which the environment is a relatively fixed entity. Although people make choices, their choice is limited to the alternatives available to them. In their design and/or management of the built environment, it is architects, planners, urban designers and other environmental designers and managers who establish the limits on people's behavioural opportunities by establishing the general context for those opportunities.

In discussing approaches to making city centres safer, this book therefore adopts a broadly environmental probabilist perspective which proposes that certain acts/crimes are more likely to happen in certain environments because more opportunities exist. If it is more likely for example that muggings and assaults will take place in pedestrian subways, then rational, risk averse individuals will try to avoid using them.

OPPORTUNITY AND MOTIVATION

As discussed in the previous section, human activity can be understood as a combination of 'sociological' motivation and 'environmental' opportunity where the environment is a set of opportunities within which people make choices. In terms of crime, therefore, there is what can be considered rational behaviour by sufficiently motivated criminals who weigh the ease of opportunity and the potential for gain. Similarly, there is rational behaviour by those who could be victims of crime and choose to avoid or manage the risk. It is necessary therefore to examine more closely the relationship between opportunity and motivation.

The presence of the opportunity is not a sufficient condition; a motivated offender is also required. It is also useful to consider a distinction between the underlying state of motivation and the 'operational' motivation. Temporary changes in the underlying state of motivation may also be provoked by, among other things, alcohol and/or drugs or anger. Equally, crime occurrence is 'not the direct, unmediated result of motivation' (Brantingham & Brantingham, 1991, p.48) and thus, as is discussed further below, criminal conduct is highly susceptible to variations in opportunity. As this book's concern is with the committal of the criminal act, it may be most constructive to follow a neo-classical utilitarian analysis of human conduct. Becker (1968, from Cornish & Clarke, 1986, p.5) argues that a useful theory of criminal behaviour 'can dispense with special theories of anomie, psychological inadequacies or inheritance of special traits and simply extend the economists' usual analysis of choice'.

A Choice Perspective on Criminal Motivation

Cornish & Clarke (1986, p.1) propose: 'a rational choice theory of criminal behaviour which views the bulk of offending as the outcome of largely reasoned decisions about the costs and benefits involved'. In essence the choice perspective argues that people calculate the expected utilities of criminal and non-criminal activities and choose the one with the greatest expected utility (Johnson & Payne, 1986, p.171). The term 'rational' in this context is somewhat problematic, particularly as 'rational' is often used as a synonym for 'reasonable', rather than – as in this case – involving conscious decision-making. Many decisions by individuals are rational to them but may be considered to be entirely unreasonable by others. For a detailed exposition of the 'rationality' of crime see Cornish & Clarke, (1986, p.1–16); for a dissenting view see Trasler (1986, p.17–24). The apparent rationality in the decision-making is not absolute but is a 'bounded' or 'limited' rationality and may, in practice, be distorted by pressures of time, emotion, alcohol or drugs. There may also be greater rationality in instrumental than in affective crime. The 'bounded rationality' hypothesis argues that behaviour is reasoned within constraints, but is not necessarily fully rational in the strict expected utility maximisation sense (Johnson & Payne, 1986, p.172). Thus, as Cornish & Clarke (1986, p.7) argue:

> decision making in the real world is only rarely the deliberate weighing of finely balanced alternatives or the careful analysis of costs and benefits that economic theory portrays. Life is too short and the decisions too many. As a result, people pay attention to only some of the facts at their disposal, they employ short cuts or rules of thumb to speed the decision process, they may perform poorly as a result of fatigue or alcohol, and under pressure of time they may make last-minute changes of plan.

The perspective is also controversial because it seems to 'depathologise' crime and suggests that offenders are more thoughtful and sophisticated than they actually are. Thus, as Cornish & Clarke (1986, p.v) suggest, it may contradict our 'deeply held and abiding fears about crime' which: 'depict it as irredeemably alien to ordinary behaviour – driven by abnormal motivations, irrational, purposeless, unpredictable, potentially violent and evil'.

Nevertheless, the perspective recognises that, first, the degree of reasoning involved will vary from offender to offender and from crime to crime, and, secondly, the possibility of pathological motives acting in concert with rational means to secure 'irrational' ends. Even in the case of offences seemingly pathologically motivated or impulsively executed, rational components are often present. Thus, without denying the existence of irrational and pathological components, Cornish & Clarke (1986, p.vi) argue that the rational components should be examined more closely. Furthermore, they argue that the tendency to 'overpathologise' offending 'hinders more constructive attempts both to analyse criminal behaviour effectively and to devise better crime-control strategies'.

Hence, in terms of the design and management of the public realm and of city centre safety, the most constructive way forward is to assume that there are elements of rationality or choice to the committal of crime; when the particular threshold of motivation in the individual at any particular moment in time coincides with the availability of opportunity, the individual then weighs the costs and benefits against the risks and effort, and chooses whether or not to commit the crime.

Decision Stages in Crime Committal

The decision stages in the criminal act are affected to different degrees by situational contingencies. Brantingham (1989) argues that there are three critical phases of decision-making in the commission of a criminal act: the initial decision phase; the search phase and the criminal act phase.

The Decision Phase

The decision phase 'may involve conscious, rational deliberation . . . or it may be essentially emotional, irrational, and instantaneous' (Brantingham, 1989, p.334). Bennett (1986) argues that the *initial* decision whether to offend or not is socially or psychologically rather than situationally determined. Similarly other decisions, such as whether to continue offending or to desist from offending, will also relate to the underlying motivation for crime.

The Search Phase

The search phase may be: 'short or protracted, but basically involves locating a potential victim in time and space' (Brantingham, 1989, p.334). Clarke (1992, p.21), however, notes that some crimes may be the result of opportunities seized, rather than ones sought or created, while Heal (1992, p.260) emphasises that it is the individual's 'learned perception' of the situation not the situation itself which is critical. During this phase, motivation is contingent on opportunity. Brantingham & Brantingham (1991, p.240) argue that the location of crimes is determined through structured search and decision processes on the part of offenders and victims. They argue that search is shaped by perceptions of environmental cues that separate good criminal opportunities from bad criminal risks and offer a multi-staged rational decision-making process model of crime site selection which relates motivation and opportunity, and also concepts of 'mobility' and 'perception' (see **Table 2.7**).

The Criminal Act Phase

The final phase is closely related to the search phase and occurs when – and if – a suitable target is located. During the search phase, opportunity and motivation function interactively: opportunity affects motivation and motivation affects opportunity. The presence or lack of opportunities will entail decisions as to whether to stop or continue searching: for example, an offender may initially decide to commit an offence as a result of the need for money but refrains from doing so because of the possibility of being seen and recorded by a CCTV system. As Bennett (1986) notes the *final* decision whether to offend against a *particular* target is situationally determined. Hence, while environmental design is unlikely to affect initial motivation, it will often affect the ease or the difficulty of the opportunity which affects the 'operational' motivation. Thus,

Table 2.7 A model for crime site selection (adapted from Brantingham & Brantingham, 1978)

1.	Individuals exist who are motivated to commit specific offences. The sources of motivation are diverse. Different etiological models or theories may appropriately be invoked to explain the motivation of different individuals or groups. The strength of such motivation varies, as does its character which varies from affective to instrumental
2.	Given the motivation of an individual to commit an offence, the actual commission of an offence is the end result of a multi-staged decision process which seeks out and identifies, within the general environment, a target or victim positioned in time and space. In the case of high affect motivation, the decision process will probably involve a minimal number of stages. In the case of high instrumental motivation, the decision process locating a target or victim may include many stages and much careful searching.
3.	The environment emits many signals, or cues, about its physical, spatial, cultural, legal and psychological characteristics. These cues can vary from generalised to detailed.
4.	An individual who is motivated to commit a crime uses cues (either learned through experience or learned through social transmission) from the environment to locate and identify targets of victims.
5.	As experiential knowledge grows, an individual who is motivated to commit a crime learns which individual cues, clusters of cues, and sequences of cues are associated with 'good' victims or targets. These cues, cue clusters, and cue sequences can be considered a template which is used in victim or target selection. Potential victims or targets are compared to the template and either rejected or accepted depending on the congruence. The process of template construction and the search process may be consciously conducted, or these processes may occur in an unconscious, cybernetic fashion so that the individual cannot articulate how they are done.
6.	Once the template is established, it becomes relatively fixed and influences future search behaviour, thereby becoming self-reinforcing.
7.	Because of the multiplicity of targets and victims, many potential crime selection templates could be constructed. But because the spatial and temporal distribution of offenders, targets, and victims is not regular, but clustered or patterned, and because human environmental perception has some universal properties, individual templates have similarities which can be identified.

a reduction in the ease of opportunity can be regarded as the functional equivalent of 'demotivating' the offender.

TACKLING CRIME

Whatever its causation, the incidence of crime and concern and fears of victimisation are a reality. The key question therefore is what can be done. Actions to ameliorate the impact of crime and the fear of crime can be focused at any or several of the various dimensions. As discussed in this chapter, interpreted in a 'black letter law' manner, the legal dimension is a technical definition of what constitutes a crime and has potential only to alter the boundary of acceptable behaviour rather than to fundamentally affect the impact of crime on people's lives (see however the discussion of 'rule setting' in Chapters 3 and 14). As it may indirectly impact upon motivation through increasing the penalties, there have been many discussions about whether penalties can be set at a sufficiently high level so as to outweigh the likely benefits of offending (see Zimrig & Hawkins, 1973; Beyleveld, 1979; Walker, 1991). Beyond their taking appropriate precautions, it is also undesirable that approaches to crime prevention should focus on victims and further circumscribe their lives. Such measures may also seem to blame the victim for becoming a victim. In terms of crime prevention, it may therefore be more constructive to focus on the motivation and/or the opportunity dimensions.

Intervention to prevent crime and/or deter criminals can occur at the decision, search or committal stage. In accordance with their views the approaches to tackling crime and criminality by the political right and the political left have differed. Nevertheless, both have generally focused on lessening the underlying criminal motivation and are intended to influence people against the decision to commit a crime.

The Corrective Approach

The political left have generally tended to support a more 'corrective' or 'welfare' approach aimed at eliminating the external motives for committing crimes. The solutions proposed tend to be concerned with programmes of social, economic and political/community development. Where incarceration is necessary, they argue that its emphasis should be on rehabilitation. As Pitts (1996, p.251) observes: 'These ideas, known variously as "welfarism" or the "rehabilitative ideal" held that . . . it would be possible to change both the behaviour and the attitudes of children and young people who broke the law'.

In the early 1970s, Martinson (1974) famously declared that 'with few and isolated exceptions, the rehabilitative efforts that have been reported so far have had no appreciable effect on recidivism'. This pessimistic 'nothing works' statement, as Muncie *et al.* (1996, p.304) argue, 'heralded the final death knell for those who believed that modern post-war societies had the capacity to rehabilitate and/or treat offenders and reduce recidivism'.

The Punitive Approach

The political right has generally tended to support more 'punitive' approaches, involving increased policing and stricter punishments by the judicial system. In the 1980s and 1990s, given the general hegemony of the political right in Britain, America and much of Europe, the law and order debate has been dominated by images of violent crime, lawlessness and a declining morality: 'authoritarianism, retribution and vindictive punishment are continually reproduced to the almost total exclusion of welfare and rehabilitative goals'

(Muncie *et al.*, 1996, p.xx). The rehabilitative approach was dismissed as a misguided attempt at 'social engineering'.

The decline of the 'rehabilitative ideal' signalled a significant change; rather than casuality and criminality, emphasis was placed upon the administration and effectiveness of the apparatus of justice and control. The conclusion drawn was that the fear of legal sanctions was the major deterrent to crime. This approach relies on the high probability that the offence will be detected and the offender swiftly apprehended and punished. The calculus for the offender is a function of *both* the likely penalty *and* the likelihood of being caught and convicted. Furthermore, if incarceration is to be an adequate deterrent, then it must be punitive: 'because it implies blame and the severity of the punishment symbolizes the degree of blame' (Muncie *et al*, 1996, p.304). The system must be retributive and offenders must receive their 'just and commensurate desserts'.

Although the emphasis of policy during the 1980s and early 1990s has been on the efficient working of the crime justice system and crime prevention through deterrence, as Muncie *et al.* (1996, p.xx) observe: 'the discourse of the New Right has . . . been significantly tempered by an apparent failure of its policies to prevent escalating crime rates'. In Britain, for example, despite tougher prison sentences almost 50% of adult prisoners and two-thirds of young prisoners were reconvicted within two years of release (McLaughlin & Muncie, 1996, p.3). Given also the continuing disenchantment with both the effectiveness and cost of welfare programmes (see for example Murray 1984; 1990b), there has been considerable pessimism about whether crime can be controlled effectively.

The Opportunity Reduction (Situational) Approach

As a result of this apparent impasse in constructively addressing criminal motivation and since crime does not happen in a vacuum (Brantingham & Brantingham, 1991, p.48), some researchers had been looking more closely at the opportunity dimension of crime. The research, in effect, focused on the relationship between opportunity and motivation and, in particular, on what Brantingham has since termed the 'search' and 'committal' phases of the criminal act. As a result, in addition to corrective and punitive approaches, a 'situational' or 'opportunity reduction' approach was developed and codified. Heal (1992, p.260) comments:

> For those supporting situational prevention, the weighting to be attached to the various elements in the criminal equation (opportunity, learned perception and propensity to commit crime) was determined, not by theoretical argument, but by pragmatism. The individual's moral code, previous learning experience and psychological and family background were, it was recognised, less amenable to intervention and change than the opportunity he or she was confronted with.

As Cornish & Clarke (1986, p.vii) argue situational prevention leaves open the question of criminal disposition: 'Regardless of how much an offender might want to commit a crime, or commit a particular crime, situational crime prevention would make it more difficult for him'. Thus, it does not require us to necessarily understand precisely what motivates any particular individual, merely to recognise that some people are criminally motivated. In this approach, the critical dimension of the criminal event becomes the opportunity. This is also the phase where the intended audience of this book has the most purchase.

CONCLUSION

This chapter has argued that crime needs to be understood in all its dimensions, and, in particular, in terms of motivation and opportunity. To make places safer, actions can be taken that modify the motivation for or disposition towards crime or that reduce the availability of opportunities for crime. Nevertheless, as discussed above, it remains difficult to know what – if anything – can be done to lessen the underlying motivation towards crime. Thus, as Hough *et al.*, (1980, p.56) argue:

> for many purposes much crime is best understood as rational action performed by fairly ordinary people acting under particular pressures and exposed to specific opportunities and situational inducements; and this crime is best prevented by manipulating opportunities and inducements.

This book therefore focuses on the opportunity dimension as the most fruitful place from which to start for constructive actions to ameliorate the impact of crime and the fear of crime by its intended audience: architects, planners, urban designers, and other environmental managers and designers. In so far as some planning and design policies can affect either actual crime or perceptions of safety, these groups have a responsibility to explore what contribution they can make. The next chapter discusses opportunity reduction approaches in detail.

3

Opportunity Reduction Approaches to Crime Prevention

The most well-developed opportunity reduction approaches to crime prevention are the Crime Prevention Through Environmental Design (CPTED) approach and the situational approach. The former was developed by C. Ray Jeffery in his 1971 book, *Crime Prevention Through Environmental Design*. Although the CPTED approach preceded the situational approach, curiously it was not the spur to its development. The situational approach has been described by Rock (1990, from Bottoms, 1990, p.43) as the 'intellectual progeny' of Ron Clarke, the head of the Research & Planning Unit of the Home Office in the late 1970s and early 1980s. The approach was the subject of two important and influential books from the Home Office Research & Planning Unit: Clarke & Mayhew's *Designing Out Crime* (1980) and Heal & Laycock's *Situational Crime Prevention: From Theory into Practice* (1986).

The CPTED approach primarily focuses on reducing property crimes rather than on personal safety. As Wekerle & Whitzman (1995, p.13) observe: 'there was no acknowledgement that violent crime prevention in urban public space might require different measures'. In addition, the approach is highly risk averse and regards all strangers as predatory. Clarke (1992, p.7) argues that situational prevention is broader than CPTED because it encompasses legal and management as well as design solutions. Although the situational approach is broader, its applicability to the safety problems of public realm also requires qualification.

URBAN PLANNING AND THE SITUATIONAL APPROACH

It is important in the context of this book to discuss some of its origins of the approaches within urban planning. Within urban planning, the key arguments were outlined in Jane Jacobs' *The Death and Life of Great American Cities* (1961) and, specifically in terms of housing, Oscar Newman's *Defensible Space: People and Design in the Violent City* (1973). Although safety had been a major concern of urban planning prior to the twentieth century in terms of defence, Modernist theorists displayed a curious lack of interest in the relationship between crime and architecture or planning except through the environmentally determinist assumption that slum clearance and comprehensive redevelopment would remove both poverty and concentrations of – and breeding grounds for – crime.

Jane Jacobs – *The Death and Life of Great American Cities*

In her seminal book *The Death and Life of Great American Cities* – although anecdotal and somewhat impressionistic – Jane Jacobs made important and plausible observations about crime committal in urban areas and about public space. The urban dimension is important as Jacobs recognised the inevitability of strangers. Indeed, as had Louis Wirth in his seminal essay *Urbanism as a Way of Life* (1938), Jacobs saw anonymity as one of the attractions of cities. A feeling of safety was, however, one of the fundamental requirements of successful urban areas: 'The bedrock of a successful city district is that a person must feel personally safe and secure on the street among all these strangers. (Jacobs, 1961, p.40).

Jacobs (1961, p.45) also noted that the social controls on public behaviour – if not crime – which could operate in smaller settlements with greater or lesser success: 'through a web of reputation, gossip, approval, disapproval, and sanctions, all of which are powerful if people know each other and word travels', would have less success in larger, more anonymous urban settings. Furthermore, she argued:

> It is futile to evade the issue of unsafe city streets . . . the streets of a city must do most of the job of handling strangers, for this is where strangers come and go. The streets must not only defend the city against predatory strangers, they must protect the many, many peaceable and well meaning strangers who use them ensuring their safety too as they pass through. Moreover, no normal person can spend his life in some artificial haven, and this includes children. Everyone must use the streets. (Jacobs, 1961, p.45).

Jacobs identified three particular themes in particular: territoriality; surveillance *and* social controls; and the presence of people. She argues that a city street equipped to handle strangers and which makes the presence of those strangers a safety asset, must have three main qualities:

> First, there must be a clear demarcation between what is public space and what is private space. Public and private spaces cannot ooze into each other as they do typically in suburban settings or in projects.
>
> Second, there must be eyes upon the street, eyes belonging to those we might call the natural proprietors of the street. The buildings on a street equipped to handle strangers and to ensure the safety of both residents and strangers must be oriented to the street. They cannot turn their backs or blank sides on it and leave it blind.
>
> And third, the sidewalk must have users on it fairly continuously, both to add to the number of effective eyes on the street and to induce people in buildings along the street to watch the sidewalks in sufficient numbers.

> Nobody enjoys sitting on a stoop or looking out a window at an empty street. Almost nobody does such a thing. Large numbers of people entertain themselves, off and on, by watching street activity. (Jacobs, 1961, p.45).

Jacobs argued that people's feelings of security were dependent on natural surveillance and that natural surveillance was enhanced in a city by vitality and a diversity of activities and various functions that create peopled places. Jacobs was particularly concerned with the underuse of certain areas and the concern that underuse promoted greater committal of criminal acts because of the lack of potential witnesses. This raises the issue of density of use and whether all places within a city centre can have a sufficient people density to render them safe or whether at certain times of the day or night, there ought to be planned 'activity corridors' which are likely to have the required density of people. Without sensitivity, however, overly planned areas may suffer from a certain sterility, artificiality and lack of spontaneity.

Oscar Newman – *Defensible Space: People and Design in the Violent City*

Oscar Newman (1973) provides a more formal framework for Jacobs' ideas in his construct of 'defensible space'. Newman conducted an empirical survey of the locations of crimes in housing projects in New York to try to identify the relationship between physical design and crime. The resulting book, *Defensible Space: People and Design in the Violent City* (1973), puts forward an alternative for restructuring urban environments, based on his interpretation of Robert Ardrey's territoriality: 'so that they can again become liveable and controlled, not by police but by a community of people, sharing a common terrain'.

Newman's ideas were very much in sympathy with those of Jacobs (1961); it is, nevertheless, a curiosity that Newman makes few references to her ideas. There are, however, at least two important distinctions: first, Jacobs' ideas are of a 'more planning' nature, while Newman's are of a 'more architectural' nature. Secondly, while Jacobs' arguments were based on observation rather than empirical research, Newman claimed that his theory had been substantiated by a meticulous research project. Unfortunately, several commentators have raised convincing doubts about the rigour of that research and the conclusions drawn (see for example Hillier, 1973, and in particular Bottoms, 1974).

From his study, Newman noted that three criteria increased the rate of crime in any residential block: first, anonymity – people living in the block did not know who their neighbours were; secondly, the general lack of surveillance within the building – lack of windows on internal corridors, hidden entrances and stairwells – making it easier for crimes to be committed unseen; thirdly, the availability of escape routes making it easier for a criminal to disappear from the scene of the crime. Newman was particularly concerned with the idea of 'anonymity' where the urban dweller no longer had a sense of belonging to a community. From this he developed his concept of 'defensible space':

> 'defensible space' is a surrogate term for a range of mechanisms, real and symbolic barriers, strongly defined areas of influence and improved

Table 3.1 The four components of defensible space (source: Newman, 1973).

TERRITORIALITY
The capacity of the physical environment to create perceived zones of territorial influence: ● mechanisms for the subdivision and articulation of areas of the residential environment intended to reinforce inhabitants in their ability to assume territorial attitudes and prerogatives. ● mechanisms for the subdivision of housing developments to define the zones of influence of particular buildings. ● mechanisms for creating boundaries which define a hierarchy of increasingly private zones - from public street to private apartment.
SURVEILLANCE
The capacity of physical design to provide surveillance opportunities for residents and their agents: mechanisms for improving the capacity of residents to casually and continually survey the non-private areas of their living environment, indoor and out. ● the glazing, lighting, and positioning of non-private areas and access paths, in buildings and out, to facilitate their surveillance by residents and formal authorities. ● the juxtaposition of activity areas in apartment interiors with exterior non-private areas to facilitate visual surveillance from within. ● the reduction in ambiguity of public and private areas in projects so as to provide focus and meaning to surveillance.
IMAGE
The capacity of design to influence the perception of a project's uniqueness, isolation, and stigma: mechanisms which neutralize the symbolic stigma of the form of housing projects, reducing the image of isolation, and the apparent vulnerability of inhabitants. ● the distinctiveness resulting from interruptions of the urban circulation pattern. ● the distinctiveness of building height, project size, materials and amenities. ● the distinctiveness of interior finishes and furnishings. ● design and life-style symbolization.
MILIEU
The influence of geographical juxtaposition with 'safe zones' on the security of adjacent areas: mechanisms of juxtaposition - the effect of location of a residential environment within a particular urban setting or adjacent to a 'safe' or 'unsafe' activity area. ● juxtaposition of residential areas with other 'safe' functional facilities: commercial, institutional, industrial and entertainment. ● juxtaposition with safe public streets.

> opportunities for surveillance, that combine and bring an environment under the control of residents. (Newman, 1973, p.3).

Newman (1973) identifies four major interrelated components of defensible space: territoriality; surveillance; image and milieu (**Table 3.1**). As Mawby (1977b, p.176) notes, however, Newman treats these categories as fundamental prerequisites and, thus, 'fails to evaluate critically the possibility that these elements might contain contradictions within themselves, and that, for example, one category might include some factors which threaten, as well as enhance, security'.

Newman's book has a tendency to reify criminals rather than to see crime as a combination of motivation and opportunity. Newman also tends to assume that crimes were committed by outsiders and appeared not to have considered whether offences might have been committed by residents. Thus, the symbolic barriers which may be effective to outsiders may be unlikely to work if the wrongdoers are insiders (Brantingham & Brantingham, 1993, p.13).

By prematurely dismissing social factors and by concentrating on physical factors, the major criticism of Newman's theory of defensible space is that it could – at best – only be partial; at worst, it might obscure the importance of other factors which might nullify attempts to make use of the theory to control crime (Davidson, 1981, p.84). Mawby (1977b), for example, argues that the theory does not take sufficient account of the presence of people or the degree of use in the space. Mawby's (1977a) study of the vandalism of public kiosks found that although visibility may protect kiosks generally, many kiosks in public situations tended to be more vandalised than those in more secluded locations. The explanation was due to the intensity of their use.

The key idea emerging from both Jacobs and Newman is the importance of natural surveillance (discussed in more detail later in the chapter). Although Newman's early works have been roundly attacked on methodological and theoretical grounds, within his work and that of Jacobs there are ideas and concepts which parallel those of the CPTED and situational approaches. In terms of the public realm, from Jacobs there is also the key notion that peopled places are generally safer places.

CRIME PREVENTION THROUGH ENVIRONMENTAL DESIGN (CPTED)

The CPTED approach was developed almost concurrently with Newman's concept of defensible space and the theories have many elements in common. The CPTED approach however extends beyond the residential context to encompass commercial sites and schools, but – significantly – not public space. The main thrust of the CPTED approach is that:

> the physical environment can be manipulated to produce behavioural effects that will reduce the incidence and fear of crime, thereby improving the quality of life. These behavioural effects can be accomplished by reducing the propensity of the physical environment to support criminal behaviour. (Crowe, 1991, p.28–29).

The three overlapping strategies of the CPTED approach are: natural access control; natural surveillance; and territorial reinforcement (**Table 3.2**).

Table 3.2 The three overlapping strategies of the Crime Prevention Through Environmental Design (CPTED) approach (source: Peel CPTED Committee, 1994).

NATURAL SURVEILLANCE ISSUES	NATURAL SURVEILLANCE OBJECTIVES
• Does the design allow us to observe?	• Design space to facilitate observation by increasing 'visual permeability', i.e. the ability to see what is ahead and around. Measure the need for privacy and/or limited sightlines against the need for personal safety.
• Is this level responsive to the needs for observation?	• Place vulnerable activities, such as cash handling/child care and other, in places that can be naturally well-monitored. Develop the potential for 'eyes on the street' by strategically aligning windows, work stations and other activity generators towards these areas of 'vulnerable activity'.
• Has the need for observation been carried consistently throughout the project?	• Take special care to ensure that each phase of the project enhances and complements natural surveillance opportunities created in the design phase. This is particularly critical with respect to the landscape and lighting phases.
NATURAL ACCESS CONTROL ISSUES	**NATURAL ACCESS CONTROL GUIDELINES**
• Does the design decrease criminal opportunity by effectively guiding and influencing movement?	• Design space to provide people with a sense of direction while giving them some natural indication as to where they are and are not allowed.
• Will safety be compromised by limiting access?	• Provide a limited number of access routes while allowing users some flexibility in movement.
• Does the design develop natural access control opportunities without considering their impact on natural surveillance?	• Take special care to ensure that natural access control opportunities enhance and complement natural surveillance objectives.
TERRITORIAL REINFORCEMENT ISSUES	**TERRITORIAL REINFORCEMENT OBJECTIVES**
• Does the design act as a catalyst for natural surveillance and access control opportunities?	• Enhance the feelings of legitimate ownership by reinforcing existing natural surveillance and natural access control strategies with additional symbolic or social ones. This might include the use of symbolic barriers or signs.
• Does the design create ambiguous space?	• Minimise the creation of ambiguous spaces (a space is ambiguous when it lacks any sort of clue as to what it is for, and who it is for). Accomplish this by identifying potential 'leftover spaces', for instance those above ground spaces between a building's underground and its property line. Then take some positive action to develop this space so that users of the property take responsibility for it.
• Will the design create heavy or unreasonable maintenance demands?	• Design space to allow for its continued use and intended purpose. Limit the need for maintenance wherever it affects natural surveillance and access control.

Access Control and Surveillance

Natural access control is aimed at decreasing opportunity through denying access to crime targets and creating a perception of risk in offenders, while natural surveillance is aimed at keeping intruders under observation. One of the CPTED approach's key elements is its emphasis on the exploitation of 'natural' forms of surveillance and access control. In this context the term 'natural' refers to deriving surveillance and access control as a result of the routine use and enjoyment of the property. Mechanical and organisational forms of access control and surveillance also form part of the CPTED approach. Access control and surveillance are discussed in more detail later in this chapter.

Territorial Reinforcement

The third design strategy, territorial reinforcement, often includes surveillance and access control strategies. Territoriality was also a key part of Newman's defensible space. Robert Ardrey (1967), in his book *The Territorial Imperative*, argued that man was a territorial animal, intent on defining and defending his personal territory. This view is disputed however and, others such as Hillier (1973), argued that the archaeological and anthropological evidence

> almost uniformally supports the social view of man – that he is a man because he is social . . . All our basic cultural facts – languages, production systems, cities above all – confirm this observation.

Hillier (1973, 1988; Hillier & Hanson, 1984; Hillier *et al.*, 1983) has been a persistent critic of the territoriality in Newman and others following in his wake, particularly Alice Coleman (1985). Hillier accepted that there is a stronger feeling of territoriality in segregated spaces and people are more likely to challenge the presence of strangers. However, he argues that this is (only) because people felt intruded upon and more unsafe: 'No one feels the need to question strangers passing down a street. On the contrary their natural presence increase the sense of security.' (Hillier, 1988). Hillier's principal criticism of Newman's and Coleman's ideas is that they advocate defensible enclaves which exclude all strangers without necessarily distinguishing between Jacobs' 'predatory strangers' and the 'many, many peaceable well-meaning strangers'. Hillier (1988) argued that the feeling of safety in public space is enhanced by the presence of people and that strangers police space and inhabitants police strangers. This should however be qualified by the observation that people generally feel safer in the presence of strangers of the same socio-economic and/or cultural group as themselves and feel uncomfortable if the strangers are of a different social class or are of certain unsavoury groups.

The CPTED approach is less insistent about the biological significance of territoriality and argues more simply that physical design 'can create or extend a sphere of influence so that users of a property develop a sense of proprietorship over it' (Peel CPTED Committee, 1994, p.3). The expression of proprietorship means that, for example:

> a clean, well-lit attractive store will present behavioural and environmental cues which tells the 'normal user' that they are safe and only accepted

behaviours will be tolerated. The same cues have an adverse effect on 'abnormal users'. The design of space and the way people are behaving will give the impression that the abnormal user will be observed, stopped or apprehended. (Peel CPTED Committee, 1994, p.4).

THE SITUATIONAL APPROACH

The situational approach aims to reduce opportunities for offending by increasing the risks and difficulties and/or by reducing the rewards of crime. Hough *et al.*, (1980) define situational crime prevention measures as:

(i) Measures directed at highly specific forms of crime;
(ii) which involve the management, design or manipulation of the immediate environment in which these crimes occur;
(iii) in as systematic and permanent a way as possible;
(iv) so as to reduce the opportunities for these crimes;
(v) as perceived by a broad range of potential offenders.

As discussed in the previous chapter, the situational approach is useful because it primarily focuses attention on the opportunity for crime. Thus, it is not necessary to understand precisely what motivates any particular individual, merely to recognise that some people are criminally motivated. The premise of the situational approach is that since crime is always a choice – not simply between illegal alternatives, but also between legal and illegal courses of action – crimes can be made so risky and difficult or so relatively unrewarding, that crime rates will fall (Clarke, 1992, p.22). This does not, however, obviate the need or necessity for further research into the motivation and decision-making of criminals. As is discussed later in this chapter, well-motivated criminals are unlikely to be deterred from carrying out the criminal act but they may change location, time, method, target or even the type of crime contemplated. Equally, the frictional effect of displacement may dissipate the criminal motivation such that no crime is committed.

Hough, Clarke & Mayhew (1980) classified situational measures into eight groups:

- target hardening;
- target removal;
- removing the means to crime;
- reducing the pay-off;
- formal surveillance;
- natural surveillance;
- surveillance by employees; and
- environmental management.

Clarke (1992, p.11–12) amended and added to these. His twelve techniques of situational crime prevention can be grouped into three sets: those that increase the offender's efforts; those that increase the offender's risks; and those that reduce the offender's reward (**Table 3.3**). The techniques are as follows:

Table 3.3 The twelve techniques of situational crime prevention (source: Clarke, 1992, p.13).

INCREASING THE EFFORT	INCREASING THE RISKS	REDUCING THE REWARDS
TARGET HARDENING • Steering locks • Bandit screens • Slug rejector device • Vandal-proofing • Tamper-proof seal	ENTRY/ EXIT SCREENS • Border searches • Baggage screening • Automatic ticket gates • Merchandise tags • Library tags • EPoS	TARGET REMOVAL • Removable car radio • Exact change fares • Cash reduction • Remove coin meters • Phonecard • Pay by check
ACCESS CONTROL • Locked gates • Fenced yards • Parking lot barriers • Entry phones • ID badges • PIN numbers	FORMAL SURVEILLANCE • Police patrols • Security guards • Informant hotlines • Burglar alarms • Red light cameras • Curfew decals	IDENTIFYING PROPERTY • Cattle branding • Property marking • Vehicle licensing • Vehicle parts marking • PIN for car radios • LOJACK
DEFLECTING OFFENDERS • Bus stop placement • Tavern location • Street closures • Litter bins • Spittoons	SURVEILLANCE BY EMPLOYEES • Bus conductors • Park attendants • Concierges • Pay phone location • Incentive schemes • CCTV systems	REMOVING INDUCEMENTS • 'Weapons effect' • Graffiti cleaning • Rapid repair • Plywood road signs • Gender-neutral phone lists • Park Camarro off street
CONTROLLING FACILITATORS • Spray-can sales • Gun control • Credit card photo • Ignition interlock • Server intervention • Caller-ID	NATURAL SURVEILLANCE • Pruning hedges • 'Eyes on the street' • Lighting bank interiors • Street lighting • Defensible space • Neighbourhood Watch	RULE SETTING • Drug-free school zone • Public park regulations • Customs declaration • Income tax returns • Hotel registration • Library check-out

Increasing the Effort

The first set of four techniques involves increasing the effort required on the part of the offender.

Target Hardening

The most obvious way of reducing criminal opportunities is to 'target harden' by increasing the degree of difficulty of the offence. This may involve, for example, the use of locks; reinforcing material screens, meshes and grilles; safes; immobilising devices, etc. Although most often applied to property, people may also target-harden themselves by for example taking self-defence classes or carrying a weapon.

There are a number of important issues concerning target hardening. First, it often leads to constraints on the use, access and enjoyment of the hardened environment. Secondly, the value of target hardening must be measured against the actual crime risk. In areas of low risk, target hardening may unnecessarily exacerbate fears of crime, and more generally is less desirable if it involves increasing levels of fear. Thirdly, it is possible that most target hardening simply leads to deflection onto another target unless most targets in the immediate vicinity are protected. Finally, as the Scottish Office circular, *Crime Prevention* (1994, para. 15) warns, target hardening should not be developed to the exclusion of other preventative strategies: 'It is important to avoid producing a fortress society where people retreat into their homes and become less interested in the safety of their neighbours and the wider community'.

Access Control

Criminal opportunities can also be reduced by controlling access. The measure forms a central component of the CPTED approach and Newman's defensible space. The intention of access control is to admit only those with legitimate purpose. Although access to many private premises is controlled, many public facilities reserve the right to practise access controls but only use it as a last resort. While controlling access to public facilities such as car parks may be appropriate, controlling access to public space is a more controversial issue. The problem arises because the public space of city centres is usually regarded as a collective or civic amenity. Thus, public space – by definition – should permit unhindered access and be accessible to all. Worpole (1992, p.5) sees the city centre as an 'important neutral territory, a site where people can mix and mingle without feeling socially embarrassed, where to some degree everybody is equal'. He also argues that 'the majority of people still feel that the town centre belongs to everyone'. Furthermore, as Reeve (1996, p.74) notes under English law, 'there is a presumption that all publicly accessible space – including where payment is charged for admission and space which is privately owned – is equally accessible and usable by everyone'.

The consideration of access controls for public spaces raises two important and related issues: first, it may entail the effective 'privatisation' of that public space, and secondly, the issue of exclusion. Although public and private space is – by definition – dichotomous, there are many quasi-public spaces where access is by permission of the owner. If access control is practised explicitly and widely, spaces cease to be 'public'. As Scruton (1982, p.375–376) notes one of the rights of private property is the ability to exclude (the others being the 'right to use' and the 'right to transfer'). To protect their investment and to ensure safety for their customers, private agencies have sought increased control over ostensibly public spaces, extending the logic of the highly-controlled environment of the shopping mall, thereby rendering them increasingly private and exclusive. The space can become increasingly geared towards consumption and only consumption. This consumption may be enhanced by or contingent upon the removal, control or displacement of groups and activities that have no commercial value. Thus, certain 'undesirable' individuals or groups, such as those deemed to be unsavoury or whose mere presence creates anxiety in others, may be excluded for the well-being and security of others.

This raises the related issue of exclusion. Access control implies an ability not to admit – and, therefore, to exclude – certain individuals or groups. The policy is usually risk averse; as it is better to be safe than sorry, it is better to exclude too many, rather than too few. Thus, the innocent may also be excluded. In addition, as Ellin (1996, p.145–6) warns, it allows 'a certain ignorance regarding social differences and therefore a fear of them, and the generation of myths, stereotypes and stigmas associated with "the other"'. On a wider scale, exclusion can lead to segregation raising important issues of social justice. Davis (1990, p.294), for example, notes how in Los Angeles traditional 'luxury enclaves', such as Beverley Hills and San Marino, have restricted access to ostensibly public facilities to 'residents only' by using municipal regulations to build 'invisible walls' in the form of preferential parking, residential requirements for access to parks, or restricted opening times (see also Chapter 10).

The issues of segregation and exclusion are controversial and raise major issues of civil liberties and individual freedoms. The particular concern is that the introduction of access control is the 'thin end of the wedge' which – unless checked – will inevitably result in the erosion of a democratic – democratic referring in this instance to universal accessibility – public realm where all people can mix freely. Nevertheless, the dwindling supply of meaningful public spaces has led to a diminution of the public realm which cities once personified. Sudjic (1996) notes how the city centre was 'once shared by every group in the community, and the exclusive preserve of none . . . In future it looks as if the city centre will become ever more narrowly divided turf.'

In practical terms, however, it is often the *degree* rather than the *fact* of exclusion that is at issue: first, who should be excluded to make it safer for the majority? Secondly, who is qualified or empowered to make such decisions? Thirdly, what checks and balances are there to ensure that the power is not abused? While some form of access control may be necessary – particularly as a last resort or the ultimate sanction – the other more managerial techniques of the situational approach may permit greater scope for action without necessarily eroding the democratic public realm.

Deflecting Offenders

Deflecting offenders involves channelling potentially harmful behaviour in more acceptable directions. As discussed later in this chapter, this is a 'benign' displacement. It involves anticipating trouble and taking steps to moderate its effects or channel it in less harmful ways. To illustrate this concept, Clarke (1992, p.14) cites an example from nineteenth century Italy. In response to a problem of people urinating in the streets, Lombroso suggested that those committing the offence should be locked up. His pupil, Ferri, suggested an alternative: the provision of public urinals. More modern examples of this may be illustrated by the techniques used to control football crowds, where the arrival and departure of supporters is scheduled to avoid the long periods of waiting around that promote trouble and, within the stadium, rival fans are physically separated to reduce the probability of fights (Clarke, 1992, p.14). Similarly scheduling the last bus to leave immediately after closing time might interfere with – or deflect – the post closing-time brawl. Litter bins and 'graffiti boards' – for people's public messages – are other examples of this measure (Clarke, 1992, p.14). This technique provides scope for the positive management of public space.

Controlling Facilitators

Exercising some form of control over the facilitators of crime (i.e., removing the means to commit crime) also reduces the opportunities for crime and, if the crime is still committed, may reduce the severity of its consequences. Clarke (1992, p.15) identifies a range of facilitators: alcohol; cars; cheques and credit cards; and telephones. One form of sanction currently being discussed in the UK is the removal of offenders' driving licences even for offences which have nothing to do with driving.[1]

Within city centres, the control of alcohol is important. Interestingly many of the measures employed have been directed towards a relaxation and liberalisation of licensing controls and/or the introduction of extra controls on public consumption. The intention is twofold: to encourage a more responsible attitude to alcohol and, secondly, to control the unsavoury presence and behaviour of drunks and winos (see Chapters 11 and 13).

As is evident from the protests of the American National Rifle Association (NRA), control of facilitators potentially involves important conflicts with personal freedoms, and it is debatable what the appropriate trade-off between liberty and security should be. In terms of firearms, however, the significantly lower rates of murder in Britain, where firearms are much less readily available than in the US, provide at least one reason to believe that effective gun controls can reduce levels of violent crime (Clarke & Mayhew, 1988; Clarke, 1992). Furthermore, as a result of the Cullen Inquiry following the shooting of a classroom of children at Dunblane, UK gun laws are set to become even more restrictive.

1 It is recognised, however, that this sanction may not necessarily deter burglars and thieves – many of whom use cars but often drive without licences or insurance. The police, however, think that its principal use will be in helping them to arrest burglars – when they cannot catch them in the act of burglary – for violating the court order prohibiting them driving. It is easier to prove that someone has violated a court order than to prove they have burgled a particular house or shop.

Increasing the Risk

The second set of four techniques involves increasing the risk incurred by the offender.

Entry/Exit Screening

As distinct from access control, the purpose of entry screening is not to exclude people but to *increase the likelihood* of detecting those who are not in conformity with entry requirements. Exit screens serve primarily to deter theft by detecting objects that should not be removed from the protected area (Clarke, 1992, p.16). Entry and exit screening should enable the distinction to be drawn between those with legitimate purpose and those without, or, in Jacobs' (1961, p.45), terms to distinguish the 'predatory strangers' from the 'many, many peaceable and well-meaning strangers'. This may relate to the possession of prohibited goods and objects (for example, firearms) or, alternatively, to possession of valid tickets and documents (Clarke, 1992, p.16). While entry/exit screening is an appropriate measure for many public facilities, its applications in terms of the public realm are limited.

Formal Surveillance/Surveillance by Employees/Natural Surveillance

The three other techniques in this set all involve increased surveillance. The deterrence effect of surveillance relies initially upon the assumption that due to increased visibility and the risk of being seen the would-be offender is discouraged from deviant behaviour. As Mayhew (1981, p.119) argues:

> People usually try to avoid being seen when committing a crime. Burglars steer clear of occupied houses, and crimes in public space tend to be committed surreptitiously (pickpocketing and shoplifting), when the victim is in some isolated spot (robbery), or when no one is around (vandalism and auto theft). Certainly there are exceptions: bank robbery or purse-snatching, for instance, rely on an element of surprise, speed, a show of force or even disguise. And drink, temper, or other precipitating factors of the kind determine the issue elsewhere. By and large, however, most offenders appear deterred by the actual or potential presence of other people.

Although the offender does not know whether or not there would be intervention – and this may be a sufficient deterrent – if the offender actually commits the offence then some form of intervention is essential. This is especially so if the individual repeats the offence and still no intervention occurs. The deterrent effect of surveillance might, therefore, be short-lived if it is not seen to result in intervention. Thus, where the offender is undeterred by merely being seen in the act, surveillance will only operate effectively where there is what the strong possibility of what Cohen & Felson (1979) term a 'capable guardian' intervening. Cohen & Felson (1979, p.589) argue that there are three minimal elements of direct-contact predatory violations: (1) motivated offenders, (2) suitable targets, and (3) the absence of capable guardians against a violation.

The lack of any one of these elements is sufficient to prevent the successful completion of the offence.

Clarke (1992, p.18) notes that despite a lack of evidence (with the exception of Painter, 1988) about the effectiveness of surveillance in preventing crime (Tien *et al.*, 1979; Ramsey, 1989), it provides much of the rationale for calls to improve street lighting (see also Chapter 7). Thus, while surveillance may lessen the fear of crime, Clarke (1992, p.18) considers its capacity to prevent crime has been overestimated. Nevertheless that many offenders avoid being seen requires some comment. Many criminals are generally risk averse and, as Mayhew (1981, p.119) speculates, may overestimate the chances of intervention. Mayhew (1981, p.119) also suggests offenders will deliberately avoid being seen by those who are familiar with and committed to defending property, persons, or the environment under threat. Furthermore, offenders can choose their moment to minimise their chances of being seen, and only part of the criminal act may be so obviously criminal as to alert onlookers that something is untoward.

There are three groups who may be capable guardians: those who provide formal surveillance; those who provide surveillance as part of their employment; and the natural surveillance provided by people going about their everyday business.

- ***Formal surveillance***

Formal surveillance is principally provided by the 'public' police and security personnel – the 'private' police – who have an explicit law enforcement function, who are trained to carry it out and whose main function is to deter potential offenders. Formal surveillance may be enhanced by electronic hardware such as closed circuit television (CCTV). A very obvious control presence can, however, become oppressive – raising fears about a 'police state' or 'Big Brother' – particularly if there are doubts about its legitimacy or if it does not enjoy public trust, confidence or respect (see Chapter 8).

The 'public' police are primarily – through force of circumstances and resourcing – a reactive force responding to incidents once they have occurred. There are also distinct differences in both the actuality and the public perception of the public and private police. Private security forces do not have the same priorities as the public police: 'Their main concern is not to bring criminals to trial but to protect the interests of the firms that hire them' (Frieden & Sagalyn, 1989, p.235). Thus, for example, private guards operating on private space can exclude visitors whose mere presence might disturb shoppers. Furthermore, as Frieden & Sagalyn (1989, p.234) note:

> civil liberty advocates point out that the guards working on private property can operate outside the legal limits imposed on the police. Unlike the police, they are under no obligations to tell suspects about their constitutional rights or to avoid random searches, since the courts have ruled that these protections are meant to restrain government but not private firms.

Although guards are not police, blurring the distinctions – for example, in terms of uniforms – is in their interests (see Garreau, 1991, p.49). There are also doubts – particularly from the public police – about the effectiveness of private security guards. In shopping malls, security guards attract criticisms from retailers for not being alert, or about their suitability for the job and appearance (Poole, 1991, p.36). Poole, however, also notes that security employees are subjected to long hours and boring duties for poor pay and suffer a non-innovative work climate. The employment of security guards is largely unregulated and even those with criminal records have been able to obtain employment, although the Home Office (1996g) is proposing to introduce a licensing system for security guards.

- ***Surveillance by employees***

Surveillance is also provided by a more heterogeneous group such as those working in public places: bus conductors, car park attendants, receptionists, caretakers and shop owners, managers, or assistants. This group has a general responsibility for the security of the property and for supervising public behaviour in the places where they work. Markus (1984, from Clarke, 1992, p.17) for example has shown that public telephones sited where they get some surveillance by employees suffer fewer attacks. The surveillance role of employees may also be enhanced by training and incentive schemes, procedures for summoning help, strategic siting of their work stations or the provision of CCTV (Clarke, 1992, p.17). In many ways, especially in terms of the ambience of the place, their presence is preferable to that of security guards as security and safety is not their primary function. An over-concentration of security guards – whose only function is enforcing security – can itself be a fear generator. What is significant is that many of these paid staff have been replaced by technology and machines or simply not replaced at all, such as unmanned station platforms.

- ***Natural surveillance***

Given that offenders will often be aware of the small risk of being seen by the first two groups, a third group – the general public – effectively provides much of the deterrent effect of surveillance. Shotland & Goodstein (1984) identify a range of 'bystander responses': taking self-protective action (i.e., running away); indirectly intervening (i.e., calling the police); directly intervening (i.e., trying to help the victim and/or apprehending the offender); and spontaneous vigilantism when bystanders not only apprehend but also punish the offender. There is however a notable – and in most cases understandable – reluctance on the part of members of the public to become involved. Mayhew (1981, p.119) offers a number of possible reasons. First, people frequently fail to notice crime taking place. Secondly, even when suspicious, observers have difficulty in deciding whether an offence is being committed and may prefer to interpret it as something more innocent. Thirdly, studies show that immediate witnesses to quite unambiguous criminal incidents are reluctant to challenge offenders or provide direct help to victims. Fourthly, the likelihood of witnesses summoning the police, especially for minor offences, is low. In addition, people often fear retaliation or are embarrassed about reporting an offender who is known to them or, indeed, may be intimidated into not reporting the incident.

In terms of urban planning, one way of creating natural surveillance is through design, which for example ensures that occupied buildings overlook car parks, and through land use controls which encourage mixed uses, mixed use developments, and the natural animation of public spaces. The intention of the latter is to keep places busy at all times by attracting in a variety of people at a range of times for different purposes. Painter (1996, p.52) notes that:

> Increased street use is a deterrent to crime because it . . . increases the perceived risks to offenders that they will be seen, recognised or interrupted by passers-by. At the same time, ordinary members of the public feel reassured in the presence of others because of the 'safety in numbers' factor.

Through the encouragement of an appropriate ambience and activity, the peopling of public places results in many safety problems being 'crowded out'. Crowds do, however, present greater opportunities for some forms of crime such as pickpocketing.

It is important to note the need for sufficient surveillance and/or for a sufficient density of people to ensure safety. Leicester City Council's *Crime Prevention by Planning and Design* (LCC/LC, 1989, p.9), for example, warns that where:

> surveillance is not possible it must be recognised that people and property are vulnerable. Fear of crime will deter people from using areas which they consider to be dangerous. Where that fear is justified, and design and planning cannot improve safety, for example, in isolated areas with no surveillance, then the danger should be recognised in the design.

Natural surveillance can also be more formalised. In the UK, there are over thirty-five different 'watch' schemes: Business Watch; Club Watch; Shop Watch; Farm Watch; Country Watch; Sheep Watch; Poacher Watch; etc. The most extensive is Neighbourhood Watch. In England and Wales there are 143,000 Neighbourhood Watch schemes; 90 per cent of the schemes see their role as being the 'eyes and ears' of the police (Home Office, 1996d). There are also more than 20,000 Street Watch schemes in England and Wales (Home Office, 1996d). Street Watch groups, in agreement with local police and residents, work out specific routes and walk their chosen area on a regular basis. Anything suspicious is to be reported to the police and they are cautioned not to interfere (see also Hetherington & Travis, 1994; Pollock, 1994).

Such surveillance schemes can, however, be seen as encouraging people to 'spy on each other' and raises the issue of the legitimacy of one individual surveying another. The trade off is often between privacy and safety. The schemes are also cautious and risk averse. The consequences of mis-identifying an individual as a 'predatory stranger' are far less severe than mis-identifying a predatory stranger as a 'peaceable and well-meaning stranger'. Thus, in the absence of complete information, surveillance can be based on prejudices about appearance and attire. Appearances may be misleading, and it is often the mere presence – rather than any actions – of certain groups which creates anxiety. Particular social groups might therefore become stigmatised and labelled; the impression inferred from general characteristics and from previous experience may not apply to the particular individual or group.

Reducing the Reward

The third set of four techniques all involve reducing the reward to the offender.

Target Removal

Target removal involves the removal of the target from the potential crime scene or, if the target itself cannot be removed, the removal of certain, more vulnerable parts of the target. Examples of successful target removal include cash reduction measures to reduce robbery, such as card rather then coin-operated telephone boxes. Target removal can also be applied to some crimes against the person. As some people run risks of victimisation in certain conditions, it would be prudent to encourage them to be aware of the risk and to take appropriate precautions to manage or avoid it. This is akin to the avoidance strategies discussed in Chapter 2.

Identifying Property

By reducing the resale value of the property, the explicit identification of property can reduce the rewards of the crime. This technique is, however, limited in its applicability to the public realm.

Removing Inducements

The opportunities for crime may also be reduced by removing the inducements to crime by for example reducing the material gain or the 'thrill' of the criminal act. Wise (1982, from Clarke, 1992, p.20) suggests a 'gentle deterrent' to vandalism by reducing inducements such as replacing metal road signs with plywood which does not 'satisfyingly "clang" when shot at'. Similarly, it has also been argued that leaving damaged items unrepaired invites further attack. This is also the essence of the Wilson & Kelling's 'broken windows' theory of crime prevention. The theory is that:

> if a window in a building is broken and left unrepaired, all the rest of the windows will soon be broken . . . Window-breaking does not necessarily occur on a large scale because some areas are inhabited by determined window-breakers, whereas others are populated by window-lovers. Rather, one unrepaired window is a signal that no one cares, and so breaking more windows costs nothing. (Wilson & Kelling, 1982).

Wilson & Kelling also argued that the failure to deal promptly with minor signs of decay in a community, such as begging or soliciting by prostitutes, can result in a quickly deteriorating situation as hardened offenders move into the area to exploit the breakdown in control (Clarke, 1992, p.36). The benefits of rapid repair are dramatically illustrated by the New York Transit Authority's success in ridding its subway cars of graffiti (Sloan-Hewitt & Kelling, 1990, in Clarke, 1992). The success was achieved by a policy of the immediate cleansing of graffiti; gratification was removed because offenders could no longer see their handiwork publicly displayed.

The technique is more problematic with regard to personal safety. Many people find the suggestion that crime should be stopped by inconveniencing the law-abiding – rather than punishing criminals – objectionable. It may, for example, appear to place an unwarranted burden on the potential victim to curtail their behaviour or freedoms. If victimised, the victim may be regarded as having been 'contributorily negligent' or is seen to be to blame for being victimised, for example, women wearing jewellery or dressed attractively may be 'asking for trouble'. In some situations, the ostensible 'contributory negligence' can be regarded as simply being female (see Jeffrey & Radford, 1984). Similarly, in terms of property, the suggestion is that small portable and valuable property, such as mobile telephones and laptop computers, invite theft. Such issues raise critical questions about the relative culpability between the crime prone and a victim who fails to take appropriate precautions. Nevertheless, crime and behaviour are dependent phenomena. There are no absolute rights or freedoms, they are all relative to context and situation. In this context, Garofalo's (1981) observation is apt: where safety cannot be guaranteed, fear of crime is functional to the extent that it encourages people to take appropriate precautions.

Rule Setting

Many public facilities – cinemas, libraries, hospitals, parks, bus and subway systems, restaurants, etc. – regulate the conduct of their clienteles, by, for example, queuing procedures, ticketing, certain limits on behaviour (i.e., no smoking), etc. In addition, most organisations regulate the conduct of their employees: they have rules about the use of the telephone for private calls, cash handling and stock control procedures. The existence and cognisance of explicit rules means that transgressors must be prepared to incur higher costs in terms of either fear of the consequences of breaking the rules or in terms of their own conscience (Clarke, 1992, p.20–21).

When setting rules, Clarke (1992, p.20) argues that the crucial factor is that any ambiguity 'will be exploited where it is to the advantage of the individual'. Thus, new rules or procedures are introduced to more accurately define acceptable and/or unacceptable conduct, thereby making it harder for offenders to 'cloud the issue' or to make excuses for their behaviour. Such rules can be oppressive, particularly if they are framed by controlling authorities without the necessity for public consultation or public accountability. Ellin (1996, p.146), for example, notes that the sign at the entrance to Universal Studio's Citywalk theme park

> warns visitors against, among other things, obscene language or gestures, noisy or boisterous behaviour, singing, playing of musical instruments, unnecessary staring, running, skating, rollerblading, bringing pets, 'non-expressive commercial activity', distributing commercial advertising, 'failing to be fully clothed', or 'sitting on the ground for more than 5 minutes'.

Nevertheless, where the space is private by paying an entrance fee and/or by merely entering the space, the user is tacitly agreeing to abide by the rules (such as not talking during a film at a cinema). If they do not or refuse to desist, they could be ejected from the space and, thereby, deprived of a resource which they value. As a consequence, the rule setting and control is consensual. If unacceptable behaviours in public places could be sufficiently defined, the same principle could be extended to public space and only acceptable behaviours would be tolerated.

Some local authorities have introduced sets of 'rules' to assist in the management of public space. Coventry for example has introduced a byelaw prohibiting the public consumption of alcohol in its city centre (see Chapter 13). In addition to an alcohol ban, Oxford is proposing another that will prohibit 'anti-social behaviour' in any street or public place. Anti-social behaviour includes spitting, urinating, defecating, touting or importuning, and the inhalation of any substance likely to cause mental or physical incapacitation (Reeve, 1996, p.76). The need is to provide a balance between freedom and restriction so that the public space remains an enjoyable amenity and for the rule setting, and subsequent enforcement, to be consensual. The difficulty is to define certain types of social incivility, such as 'unruly behaviour', in a legally rigorous and robust way in order to remove ambiguity and enable it to be enforced effectively.

LIMITATIONS OF THE OPPORTUNITY REDUCTION APPROACHES

The CPTED and situational approaches are susceptible to criticism on a number of grounds. First, that they raise concerns about the loss of freedom, manipulative control and the image of the society created. Secondly and perhaps most challenging to their effectiveness, that rather than reducing crime, such measures simply displace it. Within this section, it is also useful to discuss the concept of 'containment'. Although, these do not necessarily invalidate the approaches, they nevertheless expose some of their limitations when used in isolation and must be borne in mind.

Control and Freedom

Heidensohn (1989, p.180) dismissively labels opportunity reduction strategies as mechanistic 'engineering solutions'. Noting similar criticism, Clarke (1980, p.340) suggests the approach is seen by many as, at best, representing an oversimplified mechanistic view of human behaviour and, at worst, a 'slur on human nature'. By overlooking the interaction between motivation and opportunity, the criticism also tends to suggest that the approach is environmentally determinist, at odds with the idea of people as moral agents and suggests that people are 'hopelessly moulded' by their environment. Clarke (1980, p.340) counters these argument by asserting that the situational approach is compatible with a view of criminal behaviour as 'predominantly rational and autonomous and . . . capable of adjusting and responding to adverse consequences, anticipated or experienced'. Similarly, while recognising that the strength of individual's motivations will vary, it would be wrong to argue that:

> situational measures are only useful in respect of certain crimes, perhaps those with a strong 'opportunistic' component, while other crimes that are more deeply motivated or committed by 'hardened' offenders need to be tackled in other ways . . . all classes of crime, even those motivated by

deep anger or despair, are greatly affected by situational contingencies. (Clarke, 1992, p.21).

It is valid nevertheless that situational and CPTED measures will be *more* effective against the more opportunistic than against the more instrumental offenders. Although it is unlikely that the design of the built environment will directly affect the underlying motivation to commit crimes, it will reduce the opportunities for those with a motive. As discussed in Chapter 2, there are also various decision stages in the criminal act, each of which may – to a greater or lesser extent – be influenced by situational contingencies. Thus, although situational factors are unlikely to affect the *initial decision* to offend, they will influence the *final decision* of the individual committed to offending (Bennett, 1986).

Measures emphasising in particular target-hardening and surveillance may also evoke concerns about the image of the society being presented. As well as the spectre of the fortress city, this criticism suggests a different spectre – the 'panopticon city' (for a fuller discussion of this issue see for example Lyon, 1994). Thus, the potentially counter-productive image of the proposals must also be considered.

A further behavioural objection to some situational and CPTED measures is that they adversely affect civil liberties and infringe personal freedoms. Opportunity reduction techniques in public space should be aimed at increasing public freedom and enhancing the opportunities for people to enjoy public places. Nevertheless, it is undeniable that many of the techniques of each approach do impinge on personal freedoms (for example, access control; control of facilitators; entry/exit screening; all manner of surveillance; removing inducements and rule setting). Some techniques however such as rule setting and deflecting offenders, may, by limiting the freedoms of some, increase the freedoms of others. The debate should concern the extent to which the restrictions are considered to be worthwhile paying: for example, is the delay in boarding an aircraft worth the reduced risk of hijacking?

The regulation of public space of the city centre is not directed at curtailing the freedom of any particular individual, rather it is aimed at ensuring the safety of all individuals using the space. There should be a preference for measures that involve a positive management of public space and retain an element of choice, thereby requiring individuals to take responsibility for their actions. The intention is also to ensure that the management is consensual.

This raises two important questions: Who manages the public realm? And by what process is this management legitimised? The traditional assumption has been that the management of public realm is a local authority responsibility. As a result of its accountability to its electorate, the local authority is considered to act in the general interest. Although democratic accountability is the local authority's traditional legitimation, the political right have argued that exercising choice in the market 'provides a more realistic and immediate form of accountability than that of local government, given the low turnout and infrequent local elections' (Hill, 1994, p.199). In the late 1980s and early 1990s, Town Centre Management (TCM) partnerships in Britain (see Chapter

5) and Business Improvement Districts (BIDs) in the USA (see Chapter 12) have often become the agency responsible for city centres. By drawing in all city centre stakeholders, the TCM partnerships seek to create a sense of collective ownership and responsibility for the management of the city centre. It is, nevertheless, important to note that TCM partnerships and BIDs will often be primarily concerned with the city centre as an arena for consumption rather than as a public space (see for example Reeve, 1996). Similarly, although there may be 'armslength' accountability, it can nevertheless – especially in the US – be considered the privatisation of a public good (see for example Zukin, 1995).

Displacement

Opportunity-reducing measures inevitably pose the possibility of displacement. Being simple and direct, reducing opportunities is an attractive means of crime prevention, but the key question is how much does the problem of displacement negate its effectiveness. Commentators who emphasise the dispositional or motivational as the most important element of the crime event will tend to argue that 'bad will out' and that restricting the opportunities for crime in one location inevitably redistributes crime rather than reducing it. Cornish & Clarke (1986, p.3) refer to this as a 'hydraulic model of criminal behaviour'. They also argue that to understand displacement more fully requires further research on offenders' own explanations for their decisions and choices (Clarke & Cornish, 1987b, p.935).

The degree of displacement is likely to correlate with the availability of alternative targets and with the offender's strength of motivation: whether the incentive to commit the crime is strong enough to encourage a change of strategy once the primary intention is thwarted (Barr & Pease, 1992, p.198). Arguably, 'opportunistic' criminals are more likely to be deterred but 'professional' or instrumental criminals may be spatially displaced or resort to other means to achieve their aims. Displacement will also be affected by the type of crime and the type of opportunity reduction measure employed,

'Temporal' Displacement

The most common form of displacement is deferment where the crime is committed at a different time of day, for example, at night rather than during the day. If only by displacing some of it into another period, this form of displacement may reduce the level of crime within a given period.

'Tactical', 'Functional' or 'Method' Displacement

Offenders may also employ a different means to achieve the same end: for example, bank robbers unable to force the modern high-technology safe turn to armed robbery of cash in transit. As Davidson (1981, p.166) observes displacement of this type is indicative of the professional criminal and suggests a high level of motivation on the part of the offender towards a particular target. In some instances it may result in a more serious crime: for example, the increased security of cars has led to the phenomenon of car-jacking.

'Target' Displacement

This is where a different crime is committed. Reducing the opportunities for one method of criminal activity may result in displacement to another; for example, shoplifters may turn to vandalism, burglars to mugging. This form of displacement may be limited by the association between types of crime and the personality of offenders: burglars do not like confronting their victim, while robbers prefer confrontation to stealth (Davidson, 1981, p.166). As with displacement of method, changing the type of offence is more likely among those more highly motivated by the rewards of crime.

'Territorial', 'Spatial' or 'Place' Displacement

This is where the same crime is committed on a different target or the crime is committed in a different area. Reducing the opportunities for criminal activity in one location may simply shift the attention of offenders to similar targets elsewhere. Reppetto (1976) concluded that territorial displacement was the most likely form of displacement, particularly as a response to target hardening since all targets cannot always be protected equally. Nevertheless, in accordance with the routine activities approach, the potential for this form of displacement is limited by the highly localised nature of much criminal activity (**see Table 6.2**).

As displacement may take place in different ways, many now accept that conclusive demonstrations of absence of displacement are extremely elusive, even impossible. The inability to detect displacement does not mean that it is not present (Gabor, 1990, p.6) while, as Barr & Pease (1992, p.188) argue, the claim that displacement is total can never be precluded by research.

To further entangle the debate about displacement, many researchers have noted a positive side-effect of some situational crime prevention measures. Poyner (1988) calls it a 'complete reverse' of displacement: a reduction in crimes not directly addressed by the preventative measures. Clarke (1992, p.25) notes how this phenomenon has been given a variety of names: such as, the 'spillover' benefits (Clarke, 1989), 'free rider' effects (Miethe, 1991), and the 'halo' effect (Scherdin, 1986). He argues that some standardisation is needed and suggests the term 'diffusion of benefits', since 'the geographical and temporal connotations of this term parallel those of the competing idea of "displacement of crime"' (Clarke, 1992, p.25).

Benign and Malign Deflection

A certain amount of displacement is not necessarily a compelling argument against preventive measures. Measures applied with varying degrees of comprehensiveness will result in different levels of displacement within and between categories of crime, but the overall level of crime may not be reduced unless the measures are widely adopted. Furthermore, if only some of the targets are protected, then the risks for the remainder will grow. In the absence of universal coverage, the incidence of crime tends to swing away from richer areas and towards poorer areas – raising important questions about social justice. Although the concentration of situational measures among the society's more affluent sections is understandable, the effect is often to displace criminal activity onto the more vulnerable and it is debatable whether all areas have the resources to implement situational crime prevention measures.

Bottoms (1990, p.7) argues that it is important to ask whether the new or displaced crime is less serious, more serious, or of the same seriousness as the prevented crime. For example, displacement may be detrimental if it leads to an escalation in the means of force used, such as improved security measures by banks requiring greater use of violence by robbers (Davidson, 1981, p.167). In this respect, Barr & Pease (1992) usefully distinguish between 'benign' and 'malign' displacement. Benign displacement involves a less serious offence being committed instead of a more serious one, an act of similar seriousness being moved to a target or victim for whom the act has less damaging consequences or even a non-criminal act instead of a criminal one. Malign displacement involves a shift to a more serious offence, or to offences that have worse consequences. Although the aim is crime reduction, rather than all displacement being regarded as harmful, benign displacement is better than malign displacement.

Barr & Pease (1992, p.199) also prefer the term 'crime deflection' to crime displacement because it focuses on the more positive achievement of moving a crime from a particular target. They also suggest a concept of 'crime flux'. Rather than the more usual one-for-one consideration of displacement, crime flux focuses on the aggregate of crime movements across offenders and crime types, thereby focusing attention on the incidence and distribution of crime. Situational crime prevention measures may – or may not – lessen the amount but will always change the incidence of crime flux.

The inevitable frictional effect of displacement or deflection will serve to dissipate at least some criminal energies and motivation. Clarke (1980, p.337) argues that there must be 'geographical and temporal limits' to displacement so that a city is able to protect itself without simply displacing crime elsewhere. Furthermore, the application of one ameliorative strategy does not necessarily exclude other reforms (Hough *et al.*, 1980, p.13). To conclude this discussion of displacement, it should also be noted that foreclosure of opportunity already occurs where, through fear of victimisation, people avoid certain places and do not present themselves as possible targets.

Containment

Related to the concept of displacement is that of 'containment'. Containment reflects a belief that certain 'crimes' or 'criminal groups' can be controlled more easily if they are spatially concentrated. In some circumstances there may be a deliberate displacement without necessarily any concern as to where it is to go provided it is outside one's jurisdiction and therefore 'someone else's problem'. Alternatively there may be a deliberate displacement to a particular area. In the latter, the 'problem' may be purposefully displaced to – and then contained within – a particular location.

Barr & Pease (1992, p.207), for example, discuss the concept of 'fuse' areas. Their argument is that where an illegal trade is unlikely to be eliminated entirely, rather than a uniform distribution of the problem, there may be some advantages to concentrations. Thus, for example, unruly pubs may be useful in allowing the quiet drinker to avoid assault by going elsewhere. Their analogy with 'fuses' is drawn from electrical circuits where a weak point is deliberately introduced so that a power surge burns out that part of the circuit, the other parts of the circuit are protected from damage and it requires only a brief search for the damaged part. They argue that there are cases – admitted or otherwise – in which policing problems, such as prostitution, are controlled or reduced by having known 'fuse' areas. The fuse area permits oversight and regulation, and when public outcry becomes too great, a focused operation can be mounted to appease it. Some cities, such as Oxford, will in effect create fuse areas if – as proposed – they introduce a byelaw banning public drinking on all but a few city centre streets.

Fuse area experiments have been tried in several European cities usually by allowing drug use and distribution in certain defined and monitored areas. Christiania, to the north of the centre of Copenhagen, provides an example of this form of liberal and consensual containment. Previously a barracks, when the soldiers moved out Christiania was colonised by the young, homeless and by large numbers of people with alternative lifestyles. In the early 1970s there were confrontations with the police over drug arrests. Following a liberal approach that accepted a person's freedom to pursue an alternative lifestyle, the Government declared Christiania a 'free city' in September 1971. By relaxing their attitude to drugs in the area, the authorities saw an opportunity to contain this 'problem' geographically and to make other parts of the city centre correspondingly safer. Its continued existence has, however, always been controversial. A by-product of its idealism and the freedoms assumed by residents – and broadly tolerated by successive governments and the police – was to make it a refuge for petty criminals and 'shady' individuals from all over the city. Nevertheless, the area seems to have popular support and has performed useful functions, such as weaning heroin addicts off their habits.

Christiania is an example of containment by inducement and by a relaxation of controls within a defined area. Containment can also be an enforced ghettoisation of a social problem or social group. The policy is usually risk averse: it is better to intern too many than too few, some of the innocent along with the guilty. The original *ghetto*, created in Venice in 1516, was a means to contain the Jewish population within its own quarter of the city. The ghetto's gates were locked at night and Jews were forbidden from living elsewhere in the city. Containment may be either *de facto* – the unintentional consequence of social policy which, for example, creates 'sink estates' – or *de jure* which is likely to entail an element of compulsion and enforcement. Containment of the latter nature has occurred throughout history as a necessary means to contain outbreaks of disease and plague, hence the term *cordon sanitaire*.

Los Angeles provides a vivid modern example of an enforced containment. In Los Angeles, as in many other cities, the presence of the homeless, beggars and winos is disturbing and frightening to many people. As images and perceptions of the city are shaped accordingly, the visibility of such people has a negative impact on usage of the city and property values. As a result, city leaders have periodically proposed schemes for removing the homeless *en masse* from the city centre in order to eliminate their negative impact on downtown property values. Such schemes, however, were blocked by other council members concerned about the displacement of the homeless into their districts. The immediacy of the problem was heightened by the physical proximity of Skid Row and therefore the homeless to the central business district (Goetz, 1992, p.542).

The city therefore promoted a 'containment' policy of concentrating social services and the homeless population within the Skid Row neighbourhood, along Fifth Street east of Broadway (known as Central City East or The Nickle), coupled with an access control policy through police action to discourage indigent people from entering the CBD. The neighbourhood effectively became an outdoor poor house with considerable social costs. While largely sparing CBD users the distressing sight of the homeless, Skid Row has become a ghetto and a major problem for the city. While containment was an effective policy to serve the city's short term purposes, as the homeless population grows Skid Row threatens to spill over into the CBD. It now requires even more repressive policing or, alternatively, some other more humane solution involving review of the city's homeless policy.

CONCLUSION

This chapter has discussed and outlined the key strategies of the opportunity reduction approaches and has also outlined some of their weaknesses and limitations. It is evident, however, that planning policies and the design of the built environment can affect crime and/or perceptions of safety. It is, nevertheless, appropriate to reiterate the important caveat that design and the various opportunity reduction techniques can only create the preconditions for a safer environment: it is not a substitute for changing the conduct or reducing the underlying motivation of offending individuals. Bottoms (1990, p.7) uses an analogy with child-raising to distinguish between social and situational crime prevention measures:

> some parents might lock cupboards or drawers to prevent their children from helping themselves to loose cash, chocolates and so forth (opportunity reduction); others will prefer, at as early an age as is possible, not to lock up anything in the house but so to socialise their children that they will not steal even if the opportunities are available.

Thus, an emphasis on the situational approach does not obviate the necessity for further research into – and actions to lessen – the motivation that influences the initial and continuing decisions to offend. To address crime fully requires all its various dimensions to be tackled. This book, however, focuses on the opportunity dimension as the most fruitful place from which to start for constructive actions to ameliorate the impact of crime and the fear of victimisation by its intended audience: architects, planners, urban designers, and other environmental managers and designers.

4

The Safer City Approach

The approach advocated in this book is a Safer City Approach. It incorporates the opportunity reduction approaches outlined in the previous chapter but adds two key 'process' dimensions: first, an emphasis on partnership and, secondly, an emphasis on social justice and a safer city for all its citizens. Durkheim (1933, from Riger *et al.* 1982, p.382) suggested that 'crime unites a community by bringing residents together in outrage against violation of common norms'. In the contemporary city, crime and the concerns about public safety do not necessarily have this unifying quality. Crime often contributes to an increasing atomization of society: the Netherlands Ministry of Justice (1985, from Bottoms, 1990, p.9), for example, noted that society has become more individualistic and concluded that, in some cases, 'this individualism leads to a tendency to satisfy personal needs at the expense of others or of the community'. Harrison (1985, p.344) argues the difference in today's inner-city is that: 'the threat is diffuse and invisible and the individual stands alone in the face of it unprotected by the community'.

Such individual actions to protect oneself and one's family, however, can lead to social isolation and the loss of collective safety. This is a classic 'Prisoners' Dilemma' situation where the apparently rational, welfare-maximising actions of individuals acting in isolation unintentionally lead to lower levels of collective welfare.[1] As Tilley (1992, p.37) notes shared community safety/ crime prevention is a collective good.

1 An understanding of the Prisoners' Dilemma may be helped by a description of the situation from which it takes its name. The following is adapted from Tilley (1992, p.36–37). Two men have been arrested for illegal possession of a firearm. It is suspected – but cannot be proved – that the firearm has been used in a more serious offence, an armed robbery. The men are put into separate cells and confronted with a choice: each may confess or keep silent, but equally each does not know what the other has decided. If only one confesses, then that one will be let off but the other will be given 20 years imprisonment. If both confess, they will each be given 15 years imprisonment. If neither confesses, they will be given 2 years imprisonment each for the lesser charge of the illegal possession of a firearm.

Prisoner A's reasoning is: 'If B confesses and I don't, I get 20 years. If B and I confess, then I get 15 years. If neither of us confess, then I get 2 years. If I confess and B doesn't, then I get off scot free. As I don't know what B will do, my best strategy – whatever B does – is to confess.' As B's reasoning is identical to A's, they both confess.

The paradoxical outcome is that they both receive their third best outcome (15 years imprisonment). If they had been able to work together they would have received their second best outcome (2 years imprisonment).

To place a Safer City Approach in a broader context it is useful to examine the Safer Cities initiatives in the UK – which emphasises the partnership dimension – and the Toronto Safer Cities project – which highlights the emphasis on the more vulnerable citizens.

THE SAFER CITIES INITIATIVE

The Origins of the Safer Cities Initiative

The Safer Cities Initiative introduced in 1988 must be placed within the context of official crime prevention measures which had developed through the 1980s. In 1983, following the research-based work on situational crime prevention within the Home Office's Research and Planning Unit, a Crime Prevention Unit (CPU) was set up. The following year, a landmark was reached with the publication of an interdepartmental circular entitled *Crime Prevention* (Home Office, 1984). As well as proposing opportunity reduction measures, which 'when set against current conventional wisdom that "nothing works" carried with it an unusual note of optimism' (Laycock & Heal, 1989, p.318), one of the most important views expressed was that crime prevention 'must be given the priority it deserves and must become a responsibility of the community as a whole'. The circular argued that crime prevention was not simply a matter for the police, and that inter-agency co-operation between the police and other local agencies was of great importance in the successful delivery of effective crime prevention (Bottoms, 1990, p.3–4).

As the police are primarily a 'reactive' rather than a 'proactive' force, crime prevention also lays some responsibility on individuals and communities to make crime more difficult and, therefore, less likely (Jones, 1989, p.314). What had previously been seen as primarily a police task was now seen as something for local authorities and the general public to be actively involved in. Interestingly this official statement echoes Jacobs' views from twenty years earlier:

> the first thing to understand is that the public peace . . . is not kept primarily by the police, necessary as the police are. Its kept by an intricate, almost unconscious, network of voluntary controls and standards among the people themselves . . . No amount of police can enforce civilisation where the normal, casual enforcement of it has broken down. (Jacobs, 1961, p.41).

In early 1986, the CPU launched the *Five Towns Initiative*. In essence, this was a demonstration project for the policies of the 1984 *Crime Prevention* circular and represented the Home Office's first attempt to put a partnership approach into action. Bolton, Croydon, North Tyneside, Swansea and Wellingborough each agreed to collaborate in what were initially 18-month pilot programmes and a Crime Prevention Officer (CPO) was appointed in each town. The Home Office met the costs of the CPO, but not those of the crime prevention schemes or local running expenses. Although paid by the Home Office, the CPOs were accountable to a multi-agency steering group consisting of local authority officers and community representatives. A variety of local developments took place in these five towns. In March 1988, a similar but more ambitious scheme, the Safer Cities Programme, was launched – one of twelve separate initiatives collectively labelled Action for Cities.

In terms of related policy initiatives, 1988 also marked the launch of Crime Concern. Although ostensibly supported by private enterprise, Crime Concern was a charity substantially funded by the Home Office. It was launched as a semi-autonomous 'capacity building' agency to work with local government, the business sector and the police and to support the development of local authority and community based crime prevention. Recent emphasis has been on the prevention of youth criminality and the development of corporate strategies (Osborn & Shaftoe, 1995, p.7). A year earlier, the Government had launched a publicity campaign aimed at the general public under the slogan: 'Together We Can Crack Crime'. The Crack Crime campaign included television advertising, posters and free guidance booklets and pamphlets. Initially the campaign reflected the Government's desire to encourage individuals to take more responsibility for their own crime protection and to encourage appropriate precautions, but over time has broadened to stress partnerships and community actions (Osborn & Shaftoe, 1995, p.6).

The Safer Cities Initiative

A central idea of Safer Cities was that no formulaic solutions were to be imposed across all areas. Thus, although the overall programme was centrally administered by the Home Office's CPU, each individual Safer City Project was to be locally driven. To achieve this, three staff with knowledge of the area were recruited to administer the project under the guidance of a steering group consisting of representatives of the local authority, police, probation and voluntary organisations. There were also business and community interests and representatives from Government agencies active in the area. Care was also taken to ensure where possible that there was ethnic minority representation.

Central government funding for each Safer City was the same for each city. In addition to the salaries of Safer Cities staff and their office costs, £250,000 per annum was made available for each Safer City to support local schemes. The local steering group was able to approve applications of up to £2,000, but applications for larger sums had to be approved by the Home Office. Although initiatives were to be locally defined and driven, as Tilley (1993b, p.46–7) observes:

> the government lacked confidence in local authorities and was not about to give new responsibilities requiring additional funds to a body in which it lacked trust . . . It is likely that with a different party in power a government initiated local crime prevention programme would be subject to local democratic control.

Twenty Safer City project areas were selected from the fifty-seven urban programme authorities. In the first wave, in 1988, Safer Cities Projects were established in the following cities: Birmingham, Bradford, Coventry, Hartlepool, Lewisham, Nottingham, Rochdale, Tower Hamlets and Wolverhampton. The second wave, in 1989, included: Bristol, Hull, Islington, Salford, Sunderland,

Wandsworth and Wirral. The third and final wave in England included: Derby, Hammersmith & Fulham, Leicester and Middlesborough.

Aims of the Safer Cities Initiative

Three major aims were given for Safer Cities: to reduce crime; to lessen fear of crime; and 'to create safer cities where economic enterprise and community life can flourish'. Tilley (1993b, p.40) argues that their order also reflects priorities in the initial emphases of the programme. The approach to crime prevention implicitly – if not explicitly – advocated was the situational approach. The apparently apolitical nature of this approach was particularly agreeable to the Conservative Government; it did not, for example, attribute the blame for crime to social or economic forces. Significantly, as well as reducing the incidence of crime, the initiative also recognised the importance of dealing with the fear of crime. Research, including the British Crime Survey, had identified:

> a striking and fairly widespread mismatch between the relative risk of victimisation and fear, leading to the conclusion that strategies dealing with fear of crime may need to be addressed separately from those relating to crime itself. (Tilley, 1993b, p.45).

More curious, perhaps, was the the third aim which related to the effect of crime on enterprise and community life. Tilley (1993b, p.46) notes:

> Fear of crime not only has damaging consequences for the person suffering from it. It may also affect the community as it reduces individuals' participation in social and community activities. It is this latter connection, together with the direct commercial costs of crime, which lies behind the third aim of Safer Cities: to create safer cities where economic enterprise and community life can flourish.

It was this understanding which allowed crime prevention to be included within the Action for Cities programme. Tilley (1993b, p.46), nevertheless, sees this aim as 'largely cosmetic'; a means of 'eliciting funding, and basically a hoped-for side effect of what was essentially directed first at crime and second at fear of crime'.

Tilley (1993b, p.47) concludes that Safer Cities operated somewhere between a demonstration project and an embryonic national programme. A certain vagueness surrounded the Safer Cities initiative: it was conceived neither as a permanent programme nor as a fixed time limited one and was eventually absorbed within the Single Regeneration Budget (SRB). The removal of the bespoke funding regime for Safer Cities projects has inevitably meant that the momentum has dissipated. Tilley (1993b, p.55), for example, had presciently warned that 'the emergent Safer Cities policies and practices may well be too fragile to survive, certainly in their present form' (**Table 4.1**). Town and city centre management schemes (see Chapter 5) and partnerships however often inherited many of the ideas and schemes originally developed within the Safer Cities initiative (see also Chapter 13).

Table 4.1 Overview of strategies put in place under the Safer Cities Initiative. Tilley suggests that all other things being equal and providing that funds are available, difficulties increase going from left to right and from top to bottom (source: Tilley, 1992, p.29).

EXAMPLES OF INTERVENTIONS BY SCOPE AND LEVEL OF INTERVENTION		
LEVEL OF INTERVENTION	**PHYSICAL INTERVENTION**	**SOCIAL INTERVENTION**
FIRST LEVEL: **Conduct of new dedicated initiatives**	**Examples:** • installation/provision of door locks • window locks • fencing • blocking or creating alleyways • lighting • creation of curtilage parking • personal alarms • burglar alarms • CCTV • aspects of risk management in schools and hospitals	**Examples:** • schemes for those at risk of offending such as youth facilities or parent support groups • victim or potential victim schemes such as advice centres • offender-based programmes such as motor projects or careers advice • training for staff in public houses in dealing with violence • aspects of risk management in schools and hospitals • Neighbourhood Watch, Pub Watch, etc.
SECOND LEVEL **Incorporation into new potentially relevant initiatives**	**Examples:** • design of new housing estates • shopping complexes • public buildings such as schools, colleges & hospitals • siting of banks & post offices	**Examples:** • new school curriculum contents • management of new commercial concerns
THIRD LEVEL **Re-examination of existing patterns of practice**	**Examples:** • council repair practices for burgled properties • council policies for removal of graffiti • rubbish collection practices • provision of screechers to female employees & students	**Examples:** • school culture & management • methods of running children's homes • patterns of service delivery in health visiting • housing allocation policies • styles of policing high-crime areas • service delivery by police to crime victims • victim support for offender victims • recruitment policies for victim support • employment services to ex-offenders • policies relating to race and gender relations in public & private sector institutions

Post-Safer Cities Initiatives

In 1990, the Home Office issued a follow-up to its influential 1984 circular. The new circular – entitled *Crime Prevention: The Success of the Partnership Approach* – encouraged all local authorities to develop strategies for crime prevention and gave illustrative examples of successful partnerships. Many local authorities have safety and crime prevention projects funded through partnerships. In 1994, there was a further Home Office circular, *Planning Out Crime*, which gave advice to local authorities, developers and designers about planning considerations relating to crime prevention. Drawing upon research and practical experience, the circular recommended a number of design, security and strategic measures to be considered by planners in the development or redevelopment of housing, commercial premises and public spaces. The circular complemented the emergence of Police Architectural Liaison Officers and the police's Secured by Design initiative. The latter is a free service offered to architects, planners and developers by the police to enable them to avoid 'obvious' security pitfalls in their designs and developments.

In Britain, in November 1995, plans were announced for the creation of a national Crime Prevention Agency (CPA) – in part a restructuring of the Crime Prevention Unit (CPU) – intended to build on the partnership approach to fighting crime (Home Office, 1995b). The CPA is intended to encourage community involvement; develop new ideas to help prevent crime and reduce the fear of crime; identify what works and pass on good practice. The agency is also intended to pull together the efforts of bodies both inside and outside of Government to improve co-ordination and effectiveness. It will be managed by a board bringing together the three main national organisations involved in crime prevention – the Home Office, the Association of Chief Police Officers (ACPO) and Crime Concern. One of the agency's projects is the CCTV Challenge (see Chapter Eight).

THE TORONTO SAFE CITY PROGRAMME

Toronto in Canada illustrates how various interactive research projects into women's concerns about safety and a subsequent campaign of awareness-raising among politicians and professionals resulted in significant changes to municipal plans and policies to help bring about a safer city. The Toronto experience has been most chronicled by two of its most committed activists, Carole Whitzman and Gerda Wekerle. As a result of the campaign, the city authorities focused on municipal strategies to prevent violence against women which committed city departments to developing urban safety initiatives. A Safe City Committee was established with members drawn from community organisations and politicians and the city's Official Plan was changed to include urban safety objectives (Wekerle & Whitzman, 1995, p.9).

The developments in Toronto must also be seen against the background of feminism as it was developing through the 1970s. As Whitzman (1992, p.171) observes:

> *Take Back the Night* marches galvanised women to fight back against male violence in their communities. *Women Against Rape* groups in

several American cities advocated for judicial change, and provided self-defense information. Books such as Susan Brownmiller's *Against Our Will* brought a certain popular acceptance to treating violence against women as a means by which all men kept all women from realising their full potential. Articles in journals began to posit women's fear of violent crime in public spaces as an impediment to accessing the city's resources. From there, it was but a short step to questioning what city planners and politicians could do to help prevent public violence against women.

Whitzman (1993, p.169) argues that the 'Toronto School' of planners, academics and community activists share a common feminist analysis which consists of:

- treating 'crime' and 'fear of crime' as gendered phenomena, and public violence as part of the continuum of acts that harm women;
- relying on a participatory research and evaluation process, allowing the people most affected to define the problem and suggest solutions;
- integrating design improvements and community development.

The first principle involves a strong emphasis on and advocacy for improving the life circumstances of disadvantaged groups, in this case women. As the feminist designers' collective, Matrix (1984, p.3), observed: 'women play almost no part in making decisions about or in creating the environment. It is a *man-made* environment'. The second is an espousal of a participatory and communicative approach to planning and research, which entails the belief that those communities especially affected by public violence 'should identify the problem and suggest solutions in their own terms, instead of their experiences being transformed by "experts"' (Whitzman, 1992, p.170). The third principle was the integration of a situational and a social approach to crime prevention. Thus, changing the physical environment was 'not seen in competition with or separate from empowering individuals and communities to fight back against attack' (Whitzman, 1992, p.170).

The research and action process was itself intended to have transformative content: although the reality of women's lives was the basis of the research and action, the research and action was also intended to improve women's lives by acknowledging their reality and enhancing their skills.

The Development of the Safe City Approach

The process of instituting the concern for issues of urban safety as a fundamental consideration of urban planning and city management began in the early 1980s. In 1982, the Toronto metropolitan government – known as Metro – formed a Task Force on Public Violence against Women and Children. Although the immediate impetus was a series of rape/murders, Whitzman (1992, p.171) notes that:

> the quasi-feminist terms of reference – emphasising 'women and children' victims of crime . . . were a result of lobbying by women's advocacy groups as well as the presence of 'femocrats' in local government.

The Task Force resulted in the creation of an advocacy agency, the Metro Action Committee on Public Violence against Women and Children (METRAC), charged with ensuring that the recommendations in the Task Force's final report were implemented. These included urban design recommendations, such as improved data collection and analysis of public sexual assault sites by police, and the creation of working groups on street safety by each of the six municipalities in the metropolitan area (Whitzman, 1992, p.171).

One of the community groups on the urban design sub-committee was Women in/and Planning – an informal support network for feminist planners. In 1985, the group sponsored an interactive research project – called Women Plan Toronto – which examined what women liked, disliked, and would like to see in cities. The project's methods involved workshops with twenty-five very diverse groups, including employed women, full-time homemakers, sole support mothers, homeless women, immigrant women in support groups, native women, young women, elderly women and women with physical disabilities (Whitzman, 1992, p.171–173). In all of these group sessions, safety was a major focus of women's comments and suggestions. Although Women in/and Planning collapsed, a new group – Women Plan Toronto (WPT) – emerged. WPT was dedicated to fighting for the planning issues – including safety – that the women had identified.

The twin themes of public violence against women and women's participation in urban planning came together in the 1987 *Women in Safe Environments (WISE) Report*. This report was also the product of open-ended workshops involving a variety of women's groups and focusing on where women felt unsafe and why. Information from the workshops was supplemented by a questionnaire, distributed selectively through women's centres and other gathering points for women (Whitzman, 1992, p.173). The results helped identify particular areas of concern, including – unsurprisingly – underground garages, public transportation, and parks. Factors which made these places seem unsafe were also identified: poor lighting, sense of isolation, the existence of 'hiding spots', the presence of groups of men loitering, etc. Whitzman (1992, p.173) emphasises how 'unlike gender neutral crime research . . . these feminist research projects utilized open-ended questions; anecdotes and narrative were the basis of text, and women's suggestions'. Thus, instead of being simply assimilated into existing crime prevention theory, women's suggestions for improvement were taken as crucially important. Both of these research projects suggested that women's experience of fear ought to be the starting point for action.

The most important challenge in terms of achieving results was to transfer the findings of the research projects into the political arena and to engineer a political commitment to implementing them. A local councillor and her assistant galvanised the New Democrat Party caucus to produce a discussion document, *The Safe City Report: Municipal Strategies for Preventing Public Violence Against Women* (Whitzman, 1992, p.173). In September 1988, the report was unanimously adopted by City Council. Although the document came from a 'ginger' group of City Council, that it was able to pass relatively easily into the municipal structure was testimony to the attention raising and lobbying actions of various groups and actors prior to the November 1988 elections for the city and metropolitan councils; as Whitman (1992, p.174) concludes: 'politicians were made well aware in 1988 that votes depended on their stand on women's safety'.

The *Safe City Report* was less ambitious than the 1984 *Report of the Task Force on Public Violence Against Women and Children*. It also noted that many of the latter's recommendations had not been implemented. The *Safe City Report's* 37 recommendations still focused on urban design and planning (16 recommendations); community participation in crime prevention (6 recommendations); public transit (5 recommendations) and policing (10 recommendations). Its real strength, however, was that the recommendations were precisely targeted at the various parts of the municipal bureaucracy. Thus, for example, a number of specific recommendations were aimed at the Planning & Development Department: including putting a policy on safety into the Official Plan; reviewing policies on land use mix and open space planning from the perspective of reducing opportunities for public violence against women, hiring a planner to assist in the development of guidelines to integrate safety issues into the planning process; and training planners in women's safety issues (Whitzman, 1992, p.173). In April 1989 a newly established Safe City Committee met for the first time. The mandate for this committee was to monitor the implementation of the *Safe City Report's* recommendations and to develop further policy to enhance the safety of women.

Safety Audits

A valuable tool to develop awareness of safety issues and to inform any subsequent actions was the use of safety audits. These were developed by METRAC for use by women in tenant associations, resident groups, or informal networks to evaluate the physical safety of their surroundings. Safety audit kits were a capacity building tool for use by community and women's groups to influence the planning and design process to reflect their safety concerns. In essence, the safety audit kit was a set of questions to focus attention of what made spaces unsafe or feel unsafe. The fifteen key headings were: general impressions; lighting; sightlines, possible assault sites; isolation – eye distance; isolation – ear distance; escape routes; nearby land uses; movement predictors; signs; overall design; factors that make the place more human; maintenance; employee policies and practices; improvements (**Table 4.2**). The report *Moving Forward: Making Transit Safer for Women* was the result of a 'safety audit' of all 68 subway stops in the TTC system. In their book *Safe Cities: Guidelines for Planning, Design, and Management*, Wekerle & Whitzman (1995) outline the design principles that emerged from their work in Toronto. Despite emerging from a different research process there is a significant correspondence with the situational approach. The recommendations are largely a response to the headings in the safety audit: for example, increasing visibility; providing choices about routes; increasing natural surveillance, etc.

Table 4.2 The Toronto Safety Audit. The audit consists of a series of questions which raise awareness of safety issues in the environment and thereby empowers the users (source: abridged from Whitzman, 1992).

- **GENERAL IMPRESSIONS**

- What is your gut reaction to the place?
- What five words best explain what it is like?

- **LIGHTING**

- Is the level of lighting good enough to let you identify a face at a distance of 25 metres? Can you see everything in the back seat of a parked car?
- How even is the lighting? Are there stripes of light and darkness?
- How many lights are out?
- What proportion of lights are out?
- Are there only one or two bulbs providing light in an area? If yes, lighting may be inadequate when (not if) even one bulb burns out.
- In waiting areas like bus shelters at transit stops, does lighting put you in the spotlight? (i.e., Can anyone outside the shelter see you easily while you can't see anything outside?)

- **SIGHTLINES**

- Is it difficult to see what is up ahead because of sharp corners, walls, hills, fences, bushes or pillars? As you walk through the space, are there places someone could be hiding without your knowing it?
- Would you be able to see better if transparent materials like glass were used instead of solid materials like concrete?
- If there are surveillance cameras, are they located in the best places? Who monitors them? What happens if the person monitoring the area sees someone being harassed or assaulted?
- Could corners be angled so that it is easier to see around them?
- Should there be security mirrors to let you see around corners?

- **POSSIBLE ASSAULT SITES**

- Are there any rooms left open that should be locked up to close off possible assault sites?
- Are there any small, well-defined areas that could be used as an assault site?

- **ISOLATION - EYE DISTANCE**

- How many people are likely to be around during the day? and late at night?
- Is it easy to predict when people will be around? Could an offender be fairly sure that no potential witness will show up?
- If there were someone nearby, would they become suspicious?

- **ISOLATION - EAR DISTANCE**

- How far away is an employee who might hear an assault or a woman's scream for help?
- How far away would other people be during the day, or late at night?
- Is the area patrolled by security guards or the police? How often?
- How close is the nearest alarm or voice intercom if you needed to call for help?
- Where is the nearest telephone? Is it near enough?

- **ESCAPE ROUTES**

- How easy would it be for an offender to disappear into the surrounding neighbourhood?

- **NEARBY LAND USES**

- What is nearby land used for? Are there stores, offices, restaurants, factories, or houses nearby?
- Is there a construction site nearby? What's it like? Is lighting adequate? Is it easy to find your way around? What about other checklist items?
- Is there a sense that the land is owned and cared for by someone, or is it a desolate, abandoned place where no one would notice an assault?

Table 4.2 *(Continued)*

- **MOVEMENT PREDICTORS**

- How easy is it to predict a woman's movements?
- Are there stairways, escalators, tunnels, sidewalks, or paths that enable an assailant to predict where the victim will be in a short time?
- Could an assailant target a woman, and then wait for her at the end of a movement predictor such as a tunnel or walkway?
- What's at the end of tunnels and walkways, etc? Are there corners, recessed doorways, or bushes where someone could hide and wait for you?

- **SIGNS**

- Are there enough signs and maps so that women can find their way around easily?
- Are there signs that should be removed?

- **OVERALL DESIGN**

- Is it easy to find your way around?
- Does the place 'make sense'?
- Is it underground? How does that make you feel?
- Are there a confusing number of different levels?
- Are tunnels too long?
- Is the place too big? Too spread out?
- Do you know where the entrance is? and the exits?
- Is the overall layout like a big park - lots of open space and with buildings far away from the nearest streets?

- **FACTORS THAT MAKE THE PLACE MORE HUMAN**

- Does the place feel cared for or abandoned? Why?
- Is there graffiti? or litter? Are there signs of vandalism?
- Would things like landscaping, benches, and better design make the place more humane, and attract more people?
- Would other materials or colours improve your sense of safety?
- Are there public washrooms? If not, should there be? If yes, are they in good locations or are they hidden away down some isolated corridor? How easy is it for an intruder to get into the washrooms?

- **MAINTENANCE**

- Is the place well-maintained from the point of view of lighting, litter, broken windows, telephones, etc?
- How long does it take for something to be fixed after it is reported?
- Is good maintenance a high enough priority?

- **EMPLOYEE POLICIES & PRACTICES**

- How do employees respond to women's concerns about their safety, and to survivors of sexual assault?
- Are there enough security guards?
- Is it, in fact, employees who are harassing or assaulting women?
- Is information posted or released on sexual assaults that occur in the space?
- Does your workplace have policies on sexual assault and harassment?

- **IMPROVEMENTS**

- What improvements would you like to see?
- Do you have any specific recommendations?
- Can you think of another place that already has the positive features you are proposing?

Achievements

Prior to the 1992 municipal elections, *A Safer City: the Second Stage Report of the Safe City Committee* was prepared and passed by the Toronto City Council. One of the major achievements was the inclusion of specific commitments to the issue of urban safety in the 1993 *City of Toronto Official Plan.* The policy states:

> **7.20 Planning for a Safer City**
> It is the goal of Council to promote a City where all people can safely use public spaces, day or night, without fear of violence, and where people, including women and children and persons with special needs, are safe from violence. Accordingly, Council shall adopt development guidelines respecting issues of safety and security and shall apply these guidelines in its review of development proposals.

As Wekerle & Whitzman (1995, p.9) note the starting point for the city's crime prevention initiatives was what City Departments could accomplish within existing programmes and budgets. The Planning & Development Department for example developed training programmes for staff to increase awareness of safety and security issues; a design guidance book was published and distributed to all planning staff and to other professionals in the development and housing field. The guidelines were adopted by the City Council and applied in the development review of new projects. The changes affected urban safety in the physical environment of the city, especially in public space, and the message sent out by the City was that local government had the responsibility to ensure that public space was safe and accessible to the public (Wekerle & Whitzman, 1995, p.9).

The approach was not limited to situational methods, the City also focused on projects of social prevention. The Health Department developed an education programme for workplaces, concentrating on domestic assault. The City provided free self-defence courses during working hours for the City's female employees; developed a new programme on sexual assault and sexual harassment prevention; strengthened community action by providing funding for community groups fighting violence against women and resource kits were developed to help organise neighbourhood crime prevention (Wekerle & Whitzman, 1995, p.9).

By 1992, Whitzman (1992, p.176) was able to observe that the impact of considering womens safety was beginning to reach the various actors involved in planning and design in Toronto. The private sector had proved willing consumers of the information produced by METRAC and the Safe City Committee, particularly as the city's planners were also placing a greater emphasis on safety in development review. Community groups were learning about safety audit kits and other empowerment tools that they could use to influence the planning and design process. The challenge Whitzman (1992, p.176) identified for the feminist individuals and groups whose research and advocacy led to the changes was to ensure that the analysis of 'crime', the commitment to participatory process and an emphasis on planning and design did not get lost as the 'Toronto School' approach became part of the mainstream planning effort.

CONCLUSION: SAFER CITIES FOR ALL

A Safer City approach is reliant upon partnership between all stakeholders and vested interests. In the UK, the view expressed in official policy on crime prevention in the 1980s and 1990s has been that the police cannot do it all themselves and need the active co-operation of other statutory and voluntary agencies and of the public. In addition, there is a particular need for the main statutory agencies to pool their efforts and to work creatively together on crime prevention projects in local communities (Bottoms, 1990, p.15). For further detail of the process and techniques of partnership building, co-operation and delivery see Liddle & Gelsthorpe (1994a; 1994b; 1994c).

As the public environment of the city centre is a common property resource, the appropriate level for collective action is the city as whole. Town and city centre management partnerships are one of the forums where the various city centre stakeholders already come together to discuss issues and co-ordinate the implementation of mutually beneficial projects. From a purely commercial point of view – if nothing else – town-centres property owners are as equally locked into 'the potential for a spiral of decline as they may be party to an upward curve in profitability' (Beck & Willis, 1995, p.155).

The partnerships should also try to develop a sense of local responsibility and ownership of both the problems and the solutions. Wekerle & Whitzman (1996, p.13) argue that 'treating people as experts in the problems of their communities generates new information and solutions'. Similarly, Clarke (1992, p.28), argues that 'situational prevention starts at the bottom, by working to solve highly specific problems as experienced by particular communities or organisations'.

A Safer City Approach is an attempt to combine individual and collective crime prevention strategies to make safer cities for all. In particular, the approach should focus on those who cannot afford (or choose not) to buy individual solutions and personal security or who cannot afford (or choose not) to isolate and cocoon themselves and must (or prefer to) 'act in the world'. It has, as Wekerle & Whitzman (1995, p.13) affirm, 'a concentration on the needs of those who are most vulnerable – not only women, but older people, people with disabilities, and lower-income people'. Making a city centre safer for these groups means – in effect – making it safer for all. The value of a Safer City approach is the effectiveness of schemes to reduce the actual incidence of offences, but additionally it should also aim to change people's perceptions of their environments and be concerned with the development of a sense of control, security and confidence – and a corresponding diminution in the fear of crime – such that people are not forced to impoverish their lives through avoidance, precautionary actions and fear of crime.

5

Safer City Centres: The Role of Town Centre Management

For two hundred years, the English have been known as a nation of shopkeepers. For hundreds of years prior to Napoleon's observation and almost two centuries after, all the activity associated with shopkeeping has been conducted in the same core streets in our centres of population. It is this propensity for buying and selling, however, that currently threatens our towns and cities, causing furious debate, sometimes violent response and frequently resigned apathy among those whose livelihood is challenged by new ways of meeting consumer demand. Town centres have grown and diversified in response to wider economic and social changes. In the last twenty years, however, rapid advances in the sector have produced new forms of retailing with the expansion of modern retail developments at peripheral, edge or out-of-town locations. Combined with the decentralisation of population and increases in personal mobility, this retail diversification has created direct competition to the traditional high street. The competition is also leading to higher expectations with retailers and other city centre stakeholders having to meet customer expectations for a crime-free and nuisance-free shopping environment (Beck & Willis, 1994, p.25).

As outlined in Chapter 1, the arrival of regional shopping centres has been the greatest challenge to many city centres. Each regional centre is a purpose-built city centre-sized shopping experience under one roof with acres of free, secure car parking. Every single aspect of the buying process has been honed to perfection to extract the last 'floating' pound from a fickle buying public who flock to these facilities seven days a week, while simultaneously decrying their very existence. Despite the pronouncements of a planning regime that now recognises the damage being created by these monoliths (DoE, 1996), further similar centres will inevitably be permitted so that all major centres of population will be an easy car ride from the retailing excellence these centres promote.

Further pressure in the form of factory outlet centres and home selection using cable or satellite communications will continue to erode the available spend in the traditional high street and city centre. All this makes it all the

Copyright © 1996 Chris Hollins, Taner Oc and Steven Tiesdell more vital that a co-ordinated approach to revitalising town centres is adopted immediately if they are to survive for the enjoyment and benefit of future generations.

THE CHALLENGE FOR TOWN CENTRES

The city or town centre is regarded by many as its 'heart'; the part which provides a sense of place and belonging. URBED *et al.* (1994, p.7) note that town centres perform a wide range of functions including that of: a market place; business centre; an education and health resource; meeting place; an arts, culture and entertainment zone; a place to visit and a transport centre. Although their fundamental or primary role may be defined as a market place for retailing, Jean Carr (1990, from URBED *et al.*, 1994, p.12) argues that shopping performs a wider social role: first, as an important source of ideas and information and indeed education; secondly, as an outlet for self-expression; thirdly, while shopping is a social activity for everyone, the social interaction it provides is particularly important and valued by certain groups of people (for example, for many elderly people living on their own and for mothers with young children, shopping presents a reason for getting out of the house); fourthly the social aspect of shopping is a form of recreation.

TOWN CENTRE MANAGEMENT

It was well into the process of decanting the food offering to out-of-town superstores that the first muted tolling of alarm bells began. It was recognised that 'the economic power of out-of-town retailing lies not only with the efficient layout, controlled environment, plentiful and free parking, but also in its management structure' (Gerald Eve 1995). The majority of out-of-town complexes have their own centre managers whose role it is to liaise with individual tenants, co-ordinate strategies for managing and maintaining the common space and promote the shopping centre as a whole. It was thought that the same principles would also be applicable to town and city centres. The London Borough of Redbridge was the first to spot the techniques being practised by shopping centre managers in harmonising and maximising their retailers' collective offering under a single management regime, and to understand how they could be applied to a more diverse operation encompassing several blocks of retail units. Coupled with the earlier experience of Norwich City Council in the 1960s in building a public-private sector partnership to underwrite their ambitious pedestrianisation plans, and the growing activity in the United States and Canada through their Main Street campaigns, the concept of Town Centre Management (TCM) was born. Many of the local authorities which led the way with Town Centre Management (TCM) were the larger councils with dominant major towns acting as regional shopping centres (Rice, 1995, p.9).

From the appointment of a Borough Commercial Liaison Officer in Ilford in the mid-1980s, four further schemes followed in the next four years, spreading – albeit very slowly – the knowledge and understanding of the role of TCM. It was not until 1991, and the formation of the Association of Town Centre Management (ATCM), that the pace accelerated and within the next four years a further 125 managers were in post. Today, the concept continues to escalate with few major cities not boasting their own appointed individual or in-house team. Several dozen smaller destinations are also actively engaged in forming local working partnerships to resolve key issues that challenge them.

TCM may now be succinctly defined as the co-ordination of resources from the public and private sectors working in partnership together to address town centre related issues to improve that centre for the benefit of *all* who use it. Put simply, TCM is about effecting local solutions to local problems, managing change effectively and efficiently. Ten years of painstaking effort has enabled it to grow into an internationally recognised process of inner-urban revitalisation and regeneration that is being applied in towns and cities right across the country, almost regardless of size and scale. As shown by the prominence of Town Centre Management in the latest version of PPG 6 (DoE, 1996), central and local government acknowledge the concept as a valid mechanism, while the leaders of the commercial world are gradually warming to it as an influential adjunct to their day-to-day dealings with those who plan or manage the public realm.

INITIATING TOWN CENTRE MANAGEMENT

The TCM Partnership

Many interests are involved in town centres and a variety of public and private agencies share responsibilities for providing services and facilities to city centre users. The actual division of responsibility, however, is rarely defined and, inevitably, those responsible 'do not always pull in the same direction' (Hylton, 1989 p.12). During the last decade, widespread dissatisfaction with the type of improvements undertaken in town centres led to increased pressure from the private sector, particularly from large national multiples such as Boots The Chemists and Marks & Spencer and from other agencies, for town centres to be refocused.

In response to increasing competition, the various town and city centre interests are now working in partnerships to revitalise town centres and improve their management and maintenance. Partnerships have always been a prominent characteristic of Town Centre Management schemes, and an increasing number of local authorities are now realising the benefits of adopting a partnership approach. Although community involvement in many planning issues has been limited, the involvement of town centre users in the process of improving their town can provide a valuable source of information, generates civic pride and reduces the conflict between different interest groups. The arrival of the TCM focus has also provided an additional conduit for improved liaison between police, community and commerce, cutting through layers of red tape to allow all sides to appreciate the others' constraints and resourcing difficulties. This closer communication and co-operation has been one reason why CCTV has proliferated more quickly than might have been expected.

The TCM Partnership in Practice

From the initial first awareness of the concept in a given location, key activists drawn from the public and private sectors analyse closely the issues that their town centre faces. This detailed piece of research leads to a long-term strategy which serves to focus the efforts of all those who live and/or work in the vicinity. From these individuals can be drawn a list of resources that could be made available. The management process then facilitates and co-ordinates the matching of the identified resources to the issues to be tackled. This can range from simply janitorial, environmental improvements to the pursuit of investment to fund economic regeneration and the creation of jobs. Projects are tackled in a logical, progressive sequence starting with the small scale, 'quick fixes' that build confidence and support, then moving to the more significant major tasks that require detailed organisation and substantial commitment.

With almost 200 TCM schemes of varying degrees of sophistication, resource and application, many examples may be cited to illustrate how Town Centre Management is making a positive difference. These range from the higher end of integrated closed circuit television surveillance and the deployment of a Town Centre Manager using the facilities and influence of an in-place Safer Cities regime to roll forward the management concept, through to the simple expedient of sticking up notices to warn of pickpockets and bag-snatchers.

TCM involves a team approach to comprehensive mix of activities and improvements. A strategy typically includes a range of projects covering the economic, environmental, social and cultural needs of the town centre. Wilkinson (1993) for example identifies four key aspects of TCM:

- janitorial aspects such as environmental improvements, parking policies and cleaning strategies
- development concerns which involve planning policies and control of proposed developments
- marketing efforts including promotional activity and publicity
- entrepreneurial considerations aimed at securing activities to enhance the town centre and to aid its economic development.

The Town Centre Manager

Ultimately there comes a time when the extra demands placed on the team effort become too great for the people in it who already sustain full time employment, and specialist input is required. At this juncture the recruitment of the dedicated hours of a town centre manager is the next step. Engaged through a detailed procedure, the interviewing and selection of the manager is a critical operation to buy in the services of exactly the right individual to meet the ongoing needs of the town in the immediate and long term future. As Pritchard (1995, p.16) notes town centre managers are 'proving to be powerful evangelists of the high street's cause' and argues that their role requires 'skill, experience, charisma, and dogged determination'.

The town centre manager is the mediator between the range of interests present in a town or city centre and may co-ordinate the different departments within the local authority. The city centre manager's job is to be a catalyst, to persuade and cajole others to make their investments in ways of wider benefit. Martin Garratt, formerly Nottingham's city centre manager, for example, described his job as combining three processes: marketing, innovating and co-ordinating. Although the duties and status of individual town centre managers vary from town to town, Gilmour (1994) identifies their general role as: to develop partnerships between the public and private sectors; to assess their town's strengths and weaknesses and establish an Action Plan; to co-ordinate the more effective use of existing public resources and secure private investment; and to encourage greater community participation in particular enhancement initiatives.

SAFER CITY CENTRES AND TCM

Accepting that in essence TCM is about matching local solutions to local problems, there are a number of fundamental issues that constantly reoccur regardless of the type or size of the town or city. Access, parking, environmental services and the depth and quality of the retail offering are issues commonly identified across the country. To these must be added the problems focused around crime and the fear of crime. Without doubt, safety is a matter of concern for young and old, males and females alike. There is no social divide amongst those who suffer mentally or physically in the face of the real or perceived threat of violence.

An analysis of the business or operating plans, commercial statements of intent or council committee meeting minutes referring to the many and various local TCM initiatives in the country, would indicate that – almost without exception – there is a stated commitment to improving the safety and security of the town centre environment. Frequently this pledge is given particularly high profile; nowhere treats it as an after-thought.

During the SWOT analysis process in preparation for determining the focus of the management scheme, key 'makers and shakers' will range through the various functions of their town or city. Indeed, it is extremely likely that one of these 'makers and shakers' will be drawn from the local police service, possibly a senior serving officer. Various sources of information will be gathered: market research data on customer perceptions and requirements will be tapped; business performance will be measured; media exposure will be examined, particularly recognising the unintentional bias of the local press where bad news sells copy; reported crime statistics can be scrutinised and insurance claims assessed for the prevalent types of criminal activity. In other words, a great amount of the key information that indicates the nature and scale of local commercial and social interaction has security and safety implications and dimensions. Couple this with the high profile of Home Office initiatives, such as the Safer Cities project or the Crime Concern Initiative, and there is little wonder that crime and the fear of crime are constantly in the thoughts and deeds of those involved in the TCM process. In fact, this phenomenon of high exposure for criminal acts must go a long way to explaining the inflated perception of insecurity against the real level of recorded crime statistics. Even given the fact that only a small percentage of crime is ever reported, as discussed in Chapter 2, public perception far outstrips reality.

With the steering group or management team raising the profile of safety and security in the list of priority tasks, this naturally translates through to the day-to-day working programme of the manager appointed or the functions of that management team. Building footfall, enticing the local population to return in some depth, attracting new markets drawn from a wider catchment area, pulling in visitors and tourists, all have a direct pay-back in terms of the local economy, investment and job creation. Naturally this has a high appeal to the partners in the management process which includes not only political but also commercial interests. A sense of well-being and security rates highly in the collective public mind when decisions are made on where to spend time and money. It also colours the investment calculations of developers and financial institutions.

With such a diversity of activity there is insufficient space here to identify all of the many elements of security enhancement being considered or effected by TCM schemes. The remainder of this chapter therefore describes a few of the more successful or innovative applications in more detail.

Crime and Fear of Crime Surveys

Town centre managers need a good information base on which to work. Blackpool, for example, had little market research on which to base their particular TCM Action Plan. While recorded crime statistics and simple observation suggested that their problems centred round vandalism and violence induced by excessive alcohol abuse, confirmation was sought that the indigenous population were significantly intimidated by visiting lager louts. The town centre manager created and published his own householders' survey that was distributed via the local evening paper as a supplement. Each page of the eight page questionnaire was sponsored by a local company and spot prizes were organised as an incentive to return the completed sheets. The resulting mass of information on a variety of topics was processed and analysed by the local technical college as part of their in-kind contribution to TCM. This hard data was used to create a comprehensive local plan of campaign from which to lobby the council, as licensing agent, to exercise greater control of the tourist-based pubs and clubs.

Catering for the Car

As a body, town centre managers recognised very early in the creation of the TCM concept that although car-borne shoppers represent the 'quality spend' within the market, they are also the most fickle in their loyalty to a particular town or city centre. Easy access to free parking is one of the key factors in the ongoing success of out-of-town retailing and the mobile shopper is quite unequivocal about where he or she will visit to spend. In this respect, convenience is king. Thus, car owners have to be persuaded that access into a town centre is sufficiently easy to be worth the effort. They must also be able to park quickly, simply and safely within minimum walking distance of their target stores and the price of parking has to be exactly right. Failure in any one of these prerequisites adds to the hassle factor and increases the attractiveness of other options. Get two or more of these factors wrong and you can say good-bye to the wheels and the wallet of that particular family. This understanding explains the high priority given to safe and secure car parking.

A recent survey of shoppers' perceptions of crime and nuisance in town centres and out-of-town complexes drawn from six cities (Beck & Willis, 1995, p.61) showed that 51% of town shoppers regarded car crime as a serious problem compared with only 15% of those interviewed at out-of-town shopping complexes.

In many city centres, a great deal of time and money has been directed at improving the décor, lighting, accessibility, payment process and pricing structure of multi-storey car parks. The more progressive places, such as Nottingham, Coventry, Slough and Hemel Hempstead, have increased attended supervision and CCTV monitoring, even indulging in interpersonal skills training for associated staff so that a 'warm welcome' greets the shopper and family on arrival at the car park. The Association of Chief Police Officers (ACPO) and the Automobile Association (AA) set up the Secured Car Park scheme in 1992. The scheme is open to all car parks, whether public or private, enclosed or outdoor, surface or multi-storey and encourages owners and operators to improve standards. Standards are set for, among other things, lighting, camera surveillance, staffing, provision of lifts, and widths of entrances and exits. Car parks meeting the criteria receive either Gold or Silver Awards depending on the degree of excellence achieved. Town centre managers have been instrumental in encouraging car park operators to achieve these awards; the scheme in Nottingham, for example, was called Going for Gold.

On-site facilities and information are also becoming more commonplace in the competition to attract the affluent in their cars. Those that get it right are winning the customers back from the sterility and anonymity that typifies most edge of town 'shed' developments. Town centres are learning quickly from their rivals and can still be inventive even after copying all the tricks picked up from out-of-centre developments. Hemel Hempstead, for example, recently installed the first of several 'help posts' adjacent to the major walkways to and from car parks. These brightly coloured panels feature a prominent red call button and speaker grill. They are linked directly to the 24-hour CCTV monitoring station so that reassurance, back-up or police support can be actioned immediately as the cameras pan and tilt to pick up those who have summoned help. This visible sign, that someone cares and can help, should allay fear, day or night. In fact, it is giving the streets back to the people; a fundamental aim of those who are concerned about the well-being and vitality of town centres.

The current trend towards curbing and controlling vehicular movement in town and city centres is likely to continue as the benefits become apparent and the users of in-town facilities become accustomed to the increased safety, visibility and freedom from fumes and noise. Many cities have had extensive pedestrianisation programmes. During the 1970s and 1980s, pedestrianisation of central areas was a popular initiative and often the centrepiece of town centre improvement schemes involving the removal of traffic to form pedestrian-only streets. Such schemes aimed to provide safer – in terms of people-car interaction – and more attractive shopping environments. Restricting all car access over large areas and for long periods, however, has created deserted and often unsafe areas in the hearts of cities out of shopping hours. Without the natural surveillance from the occasional passing motorist, those so inclined can commit offences with impunity: mugging, rape, burglary and ram-raiding. Newcastle's CCTV system has captured horrendous footage of acts of violence in the pedestrianised core of the city. The new emphasis is on pedestrian-priority schemes, such as widening pavements, rather than pedestrian-only schemes.

Retail Units

The retail scene has been the focus for many security-based initiatives in the attempt to drive down public perception of lawlessness and petty theft which may dominate the shopping experience. Many are created with the strong acquiescence of the major national multiples who suffer considerable reduction in profits at the hands of both professional and amateur shoplifters. In-store video camera surveillance is frequently augmented with undercover detective facilities. There has also been a strong move towards uniformed security guards on the sales floor to supplement electronic tagging systems for more expensive merchandise. There is also an increasing awareness that pooling resource and co-operation across a broad security front can be an effective deterrent against organised crime. Inter-linked radio paging systems have been developed that have overcome some of the fundamental flaws that marred earlier versions and are now available in sufficient depth to conclusively prove their value.

Falkirk was one of the first towns to invest in counterfeit note detection equipment on the back of their TCM scheme. This was an early success in the saga there to create and sustain a 'Business Watch' campaign. A network of street representatives was used to communicate to small independent traders the information and skills gained by the better equipped and trained national companies. Seminars and practical demonstration sessions coupled with specialist videos and external speakers were co-ordinated by crime prevention officers from the local police force. The success of this initially high profile set-up meant that it extended naturally and easily into Pub Watch and Crime Watch operations that are now commonplace.

A particular problem for town centre managers is security shutters on shops (**Figure 5.1**). One survey of the public showed that nearly 58% of people saw them as 'ugly but necessary', 22% thought they looked 'ugly', while only 11% did not notice them (Beck & WIllis, 1995, p.203). Many town and city centre retailers are increasingly keen to erect security shutters to prevent burglary, ram-raids and criminal damage. The shutters come in various types ranging from open grille to solid and can be fixed internally or externally. Many

Figure 5.1 External shutters on shops in Coventry city centre. A high spatial concentration of shutters can be seen as an indicator of a high crime environment, heightening the fear of potential victimisation.

insurance companies offer a lower premium for retailers who install shutters. Although security shutters are most typically installed on the windows of off-licences, hi-fi stores and jewellers – where small, relatively high-cost items that can be sold on easily can be stolen – they are becoming an increasingly ubiquitous feature of many shopping streets. The shutters are generally erected out-of-hours and at night. In some instances, they may be in place during the day. While the retailer's concern is understandable, the effect of shutters can be a fear generator. The 1994 circular, *Crime Prevention*, (DoE, 1994, p.6) noted that:

> The creation of a fortress-like atmosphere can be self-defeating. Solid roller shutters can have an adverse environmental effect, giving an area a 'dead' appearance and contributing towards the creation of a hostile atmosphere. They are also vulnerable to graffiti. This not only gives out signals about an area's vulnerability to crime, but can also deter the public from using such locations, thus losing the benefit of passive surveillance.

Nelson (1996) conducting research in three city centres found that many shoppers also complained that shutters prevented light being shed from the shops onto the streets, so reducing levels of lighting and preventing window shopping out-of-hours. Findings from a study by Beck & Willis (1995, p.204) suggest that both retailers and shoppers are resigned to their installation: 'perceived need and presumed effectiveness of shutters outweighs any aesthetic reservations. Shoppers seem to be fatalistically resigned to the use of security shutters.' Some cities, however, are actively trying to address this situation. Cardiff City Council, for example, through a grant-aided scheme is encouraging the removal of solid security shutters from shop fronts. The external shutters are replaced by internal grilles, the stall-risers are raised, and other, less obtrusive security devices are used (Nelson, 1996, p.196). Raising the stall risers and building them of brick or concrete also provides protection against ram-raiding.

Where premises are particularly unattractive – either physically or economically – they may remain void and often neglected. By attracting daubing, posters and, sometimes, an illegal shop squat, they may quickly blight the street. Empty units are always the focus of attention from the rest of the retail community around them, provoking demands for the often absentee landlords or estate agents to do something about them. Many places have put pressure on owners to renovate or clean up the facades of their buildings and premises. Nottingham, for example, ran an extensive drive to keep a uniformly high standard of shop presentation, cleaning up old units and insisting on landlord co-operation to help rent such units to new tenants as soon as possible. This has paid off with the percentage of void retail units well below the national average due to a high profile for the city, its management regime and progressive council policies.

The creation of high-density retailing, which formerly required considerable stock storage and staff support facilities as well as a large sales floor, has meant that the upper floors of many retail premises, once needed for warehousing and staff accommodation, are now vacant. Innovative and sophisticated stock control has also made much of this space redundant. The tradition of 'living over the shop' has largely ended and, despite efforts to revive it, very few cities boast a substantial number of people living in the inner area. This means that the upper floors of most retail units are deserted with blank, lifeless windows staring out. Out of shopping hours, the street loses the advantage of natural surveillance that shoppers and residents would provide and becomes dead and soul-less.

In response to the increasing amount of vacant space above retail units, the Living Over The Shop (LOTS) campaign to re-introduce residential accommodation into vacant premises and especially the floors above trading units is gathering momentum. It has also been adopted by several large national organisations, including Boots The Chemists and the National Westminster Bank group. Glasgow, Salford, Ashton and Stafford are good examples of where refurbished flats and town centre housing have been re-established, building up a ready-made workforce for the retail sector and, of course, raising the level of security and civic pride in the vicinity of the traditional high street.

Combating Physical Incivilities

Efforts on the environmental front have naturally had a high profile throughout the various management schemes. The demands of small business concerns and the more vociferous elements of the community have placed a high emphasis on raising of standards of local housekeeping. The removal of overnight flyposting has been a high priority in Wolverhampton where it was recognised that routine elimination before it is seen by the public dramatically reduces poster effectiveness. It also sends a clear signal to those who commission such activity that their efforts are not appreciated and will not succeed. This has been reinforced in some locations, such as the City of Westminster, by prosecution and in the imposition of repeated heavy fines. Businesses have been challenged to keep private property clear of posters and scrawl. In Halifax and Dundee, among many places, a professional firm of specialists has led the anti-graffiti campaign. While raising the firm's own profile, it also plays an important role in boosting civic awareness of this issue.

Combating flyposting and graffiti is increasingly frequently the responsibility of a specially recruited town centre caretaker or handyman. This role – frequently privately sponsored – has a high public profile, and is usually to be seen patrolling the streets early in the morning and at busy times of the day attending to the immediately obvious rashes of overnight mischief. This daily exercise is conducted before the regular council cleansing team starts to function and is a highly visible message to the general public that somebody cares enough to tackle anti-social behaviour. The most recent recruit to this post is the Derby Ranger; the first deployment in that city to sustain the quality and appearance of their recently pedestrianised core. Initiatives of this nature accord with what Wilson & Kelling (1982) termed the 'broken windows' theory (see Chapter 3).

Nevertheless, as Nick Falk (1991, p.16) warns, too often environmental improvements are 'cosmetic measures which fail to put new life into the heart of the town'. Baldock (1994, p.13), however, argues that environmental improvements will only make a significant contribution to the quality of life in town centres, if the other – more functional – needs of its users have been satisfied. Thus, if people cannot get the goods or services they need, or feel insecure and unsafe, the quality of the environment will be immaterial to them. Rather than environmental improvements, shoppers tend to want more or better shops, improvements to accessibility and safety before anything else.

Designing Out Crime

'Designing out crime' and the situational approach have also come to the forefront in today's planning circles. Whilst prodigious green planting, terracing, underpasses and narrow concrete pathways were popular inclusions in the 1960s and 1970s, heightened awareness of safety issues means that greater attention is paid to this in the architectural design phase. Coventry and Birmingham have reviewed many aspects of the street layout with improved security in mind. The latter is also well into a programme of underpass replacement. Lighting is a very obvious advantage when correctly installed and maintained (see also Chapter 7). In Leicester, particularly vulnerable underpasses to the university have been substantially better-lit and observed by CCTV. Monitors from the cameras play to nearby densely-used streets, so that pedestrians there can informally police the subways.

Key Performance Indicators

Most schemes seek to verify the benefits of TCM activity over the long term. This can be done by defining a number of performance indicators that can be used to monitor the progress of the management regime, allied to the general elevation in local standards and the improvement in the return on investment made by commercial companies trading in the town.

Key Performance Indicators (KPIs) are the subject of several pieces of ongoing research which should lead to a universal range of common criteria. In future these will be applied across the entire spectrum of towns and cities, enabling benchmarking and modelling to be conducted in the pursuit of greater vitality and vibrancy across the board. Examples of KPIs include: market research; footfall surveys; recording the decline in the redundant; vacant premises available to the market; the reduction in charity shop density; Zone A rental values; and, of course, the evidence of officially recorded crime statistics to demonstrate an improving situation. The latter parameter is already widely accepted as a valid mechanism for checking the improvement in the 'health and well-being' of a centre. In Nottingham, positive promotion of the reducing crime statistics has been used to increase the confidence of both residents and visitors.

In fact, the Business Plan of the Nottingham City Centre Management scheme – one of the foremost TCM operations in the country – identifies eight principal objectives in its overarching strategy (Nottingham City Centre Management, 1993). The fourth of these focuses on crime levels, pledging to monitor them and to introduce a course of action to reduce crime and the fear of crime in the city centre. This became a carefully structured plan to decrease incidents of violence during the evening, increase leisure-based aspects of city life, displace vagrants, control pamphleteers, create a day-time Shop Watch scheme and to enhance the on-street café facilities to inject a more leisurely, relaxed atmosphere into the core streets. Other related initiatives on flyposting, graffiti, car based crime and security training have been built into the overall management strategy to give a multi-faceted approach that has clearly paid dividends for Nottingham in the past three years (see Chapter 13).

CONCLUSION

Personal safety – and the feeling of security – is a prequisite of a vital and viable city centre. If a city centre is not perceived to be safe, then those people with the luxury of choice may choose not to use it, making it less safe for those who have fewer choices. The TCM Partnership and the office of the town or city centre managers are important forums where the issue of – and fears about – safety in city centres can be addressed and positive actions taken. Town Centre Management will shortly complete its first decade. During this time it has proved that it is not a short term 'quick-fix' but an imaginative approach to the process of managing change for our major – and not so major – centres of population. As towns and cities need to evolve rapidly to meet the challenge of differing retail formats and other demands on increasing leisure time, it is essential that all available local resources are mobilised to elevate quality and vibrancy. It is vital that commercial common sense and influence are projected into the local government process because of the direct relevance of both sectors to employment and wealth creation. And it is essential that we all feel safe and secure in an environment which we share with past and future generations as well as each other.

6

Policing for Safer City Centres

At the end of the twentieth century, we have seen important changes occur in the concept and use of city centres and we are already faced with critical decisions about their future. Although cities are the centre of government, the focus of intellectual and artistic leadership, they are now cursed by a confusion of purpose and a fragmentation of intention. In Britain, as in America, heritage and culture, entertainment and fashion, consumer needs and government have focused exclusively on cities. The present generation is the first in which people – liberated by the proliferation and ready accessibility of cars – have had the option if they wished to reject the city. The consequences of this rejection, however, are now becoming obvious. Britain is closely following America's example, and, although in many ways the culture is different, the comparisons are alarmingly close.

Surprisingly, a large proportion of problems suffered by urban America have been caused both directly and indirectly through neglecting city centres in planning decisions relating to commercial enterprise. Technology has enabled businesses to dispense with the city. Furthermore, the exodus to the suburbs has, in general, been by the more affluent in society. Thus, as the less wealthy pay less tax, the tax base of city centres has also diminished. British and American profiles of rising urban crime mirror each other. As the greater part of law abiding society has now abandoned cities – only returning to them for brief periods of the day to work and then to quickly depart again before dark – cities, particularly inner cities, are becoming ghettos that are enormous generators of criminal activity. Whereas the 'haves' retreat to the safety of affluent suburbia, the 'have nots' are trapped within a tightening spiral of problems.

If social order is a vital component in a well-functioning city, then the practical, visible face of such control – for the time being – is the police service, with its central duty being the maintenance of law and order. Later we will show how that dominance is rapidly eroding towards a mixture of police, private security and 'urban ranger'-style, council-funded, 'second division' law enforcement. Just as the use of cities has changed in the last thirty years, so too has policing.

Copyright © 1997 Stephen Lister and Robert Poole

This chapter examines the following issues: the effectiveness of policing strategy for city centres; how the police gain a greater understanding of patterns of crime and criminal intelligence; how the modern police service embraces concepts of Crime Prevention Through Environmental Design (CPTED) and turns them into a practical reality; how partnerships are important in crime reduction; and the use of technology. Weaknesses, within the police service, within a goal-orientated, quantitatively-measured environment will be explained, as well as the unique opportunities it possesses to work towards the ultimate aim of the provision of safe and secure city centres.

POLICE ORGANISATION

Central to the understanding of the police role in safety and security is an awareness of the fact that there are fifty-two police forces in Great Britain. They are required, to a degree, to speak with a common voice but the central control that exists is supplemented with a strong local control through police authorities giving chief constables the autonomy to interpret orders from the centre. These arrive by way of Government directives and Acts of Parliament, in the light of their own circumstances and particular priorities. Although senior police managers maintain that the service has always been run in an orderly and accountable fashion, with the election of the Conservative Government in 1979 a concerted effort began towards requiring all public services to adopt the sound principles of business practice. In 1983, the Home Office issued a landmark order, Circular 114/83, *Manpower Effectiveness and Efficiency in the Police Service*, which specifically spelt out how future policing operations should be managed:

- with a statement of the present position,
- a pronouncement of the desired position,
- the steps to be taken towards that goal with a stated time scale,
- an evaluation of accomplishments, including costs.

'Policing by Objectives' was a philosophy that demanded careful, well-managed approaches and steered policing towards activities that could be easily counted and which produced outcomes that could be measured – and more importantly attributed to the prescribed strategies (Butler, 1992).

The Police Act 1964, as amended by the Police and Magistrates Courts Act (1994), requires policing plans to reflect key national objectives set by the Home Secretary and to contain any local objectives set by the relevant police authority. The authority determines these local objectives after consulting with the chief constable and considering views expressed through police community liaison committees. Most police forces have published 'statements of purpose' to demonstrate their commitment to their community and employees. They have created corporate plans, formulated strategic plans and produced service level agreements. Some have received formal recognition in the form of Charter Mark, Investors in People, Citizens Charter and Crystal Mark. Some have ignored writing in full about what they hope to achieve but have, none the less, set about doing it.

One of the outcomes of the move to more empowered local management teams is of business processes being devolved to police divisions. This has been extended by locally-costed policing plans. Both are compounded by widely differing needs in policing priorities. The demands of national objectives still need to be met, as translated locally into modified priorities. The move to performance management and a more outcomes-driven service will require changes at all levels, involving different approaches to resource allocation, utilisation and targeting.

MAKING LESS GO FURTHER

As the police service moves towards a more accountable style of management, with the focus on output results, the implications for changes in organisation culture will magnify the effect of the change process at local level. These effects need to be built into the planning process, which needs to take account of organisational change as well as administrative and procedural changes. Each police force has seen major increases in crime in the last two decades without proportional increases in human resources. This was recognised by the Audit Commission in their 1993 report which stated: 'Demand from the public for police services has increased at a faster rate than have police resources'. In the 1994/95 Annual Report of the Humberside Police, the chief constable stated: 'Reported crime has almost quadrupled since 1974 yet staff numbers have increased by just 6%'.

At national, force and local level there is a need to generate income to supplement funding. Whilst this has been foreign to the police service in the past, it is now a requirement as an accepted way forward. The experience, knowledge and expertise of the police should be put to beneficial effect by harnessing it to more progressive ways of planning or designing out crime in city centres. Some forces are already doing this.

Police priorities put reactive needs first; proactive initiatives will always take second place. With inadequate resources it can be difficult or, sometimes, impossible to carry out proactive initiatives. However, if successfully planned and implemented they can, very often, reduce the need for reactive policing later. Planning successful initiatives to make safer cities is just one of the proactive approaches that may suffer due to increasing demands of the service. The difficulties in measuring the success of crime prevention initiatives may, however, cause police managers to seek sanctuary in more traditional policing methods which lend themselves more readily to measurement. In addition to this, rising crime in areas of burglary, car crime, property crime, fraud, terrorism, drugs, and the like, not to mention public nuisance offences, make the allocation of scarce resources increasingly difficult.

Whereas the results of public opinion surveys often show that a more visible police presence is desired (Honess & Charman, 1992), they express concerns about the increase of less serious incidents of public nuisance rather than, what is generally expected, those crimes of rape, robbery and similar serious occurrences.

POLICE ACCOUNTABILITY

The police have become far more accountable in the last decade. Apart from the usual form of inspection by Her Majesty's Inspector of Constabulary (HMIC) they are subjected to public scrutiny from many organisations, such as the Audit Commission, National Audit Office, Parliamentary Committees, Police Complaints Authority, Police Authority, lay visitors, academic institutions, public enquiries, the judicial system and the media. As protection, police executives and managers may adopt measures which help reduce the potential for criticism – sometimes to the detriment of initiatives that may be seen as proactive or that produce no measurable outcome.

The whole aspect of service performance is extremely difficult to measure effectively for an organisation like the police. For example, a police division where police regularly arrest or report many offenders may be viewed as being very efficient. A neighbouring area may have fewer offenders arrested or reported and the police may be criticised for their lack of zeal. In reality it could be that the high arrest area is being policed reactively, with crimes occurring in high numbers, whereas the low arrest area is being policed proactively and professionally and therefore the opportunity for crime to occur is minimised. Crime prevention and reduction methods regularly go unnoticed and unrewarded, unless they are dramatic and overplayed. The police must be able to make informed judgements and decisions based on priorities prevailing at any time and place to satisfy the needs of others, as well as remaining impartial and objective – and at the same time act within the law and expectations.

POLICE OFFICERS WANT SAFER CITY CENTRES

Police officers want safer city centres, it makes their job easier. They do not get paid on performance even though that is their goal. They want their families and friends to be safe, as well as the communities they serve. Safety and a tranquil environment mean peace and prosperity for all. The policy statement of the Association of Chief Police Officers (ACPO) of June 1990 states: 'The primary objective of the police service is the prevention of crime. It is to this aim that the efforts of all police officers are directed.' Indeed the best practitioner of crime prevention is an operational police officer carrying out normal patrol duties, identifying opportunities for crime and reducing them by advice or positive action. Nevertheless, there are no specific requirements for businesses or individuals to comply with universal standards for crime prevention.

In order to be effective in preventing crime, expert knowledge is required. Most police forces have a specialist department with responsibility for crime prevention or crime reduction. Training and support is given by the Home Office Crime Prevention Centre based at Easingwold, North Yorkshire. The centre ensures a consistent approach to specialist training for a small, select group of officers. Only a few forces have personnel nominated as Architectural Liaison Officers. This job requires skills which only a minute fraction of officers in the police service possess. Most forces have statistical analysts and crime managers and some have officers or civilian employees as crime analysts. The Home Office also provide services and support for crime prevention and reduction, mainly in the form of research through the Police Research Group, Research & Planning Unit and the Crime Prevention Agency (formerly the Crime Prevention Unit). ACPO provide committees and working groups that add direction to the subject, being either specific or general. Her Majesty's Inspectorate of Constabulary identify and promulgate good practice during the course of force inspections.

One of the Home Secretary's key national objectives is: 'To target and prevent crimes which are a particular problem, including drug-related criminality, in partnership with the public and other local agencies' (Howard, 1995). The majority of forces are adopting methods to ensure that partnerships to tackle crime are enthusiastically encouraged. Whilst the police have long been involved in partnership approaches in a variety of initiatives, such as Neighbourhood Watch, Crime Prevention Panels, Safer Cities, City Challenge and local projects, the recent encouragement to cement even greater partnerships will benefit everyone. Crime, and its prevention, is the responsibility of all sections of the community, not solely the police. The business sector suffers directly by lower profits due to crime losses, preventative measures and lower turnover or lower opportunity from victimised or fearful customers. Local authorities suffer revenue losses in the exodus from city centres by retailers, businesses, residents and shoppers. Insurance premiums increase due to the escalating cost of crime and affect us all.

Crime prevention should be one strand of a number of design and management strategies to keep areas 'crime-free'. It is in everyone's interest to work together, as individual initiatives alone are often unable to influence the greater environment. Whilst the police have, in the past, usually taken a lead role in stimulating partnerships, now is the time for other sectors of the community to take that lead and include a willing police service as a valuable partnership member. Partnerships can become a most powerful weapon against criminality.

LEARNING FROM REPORTED CRIME

Making sense of criminal acts by the analysis of crime statistics can be akin to trying to complete a jigsaw puzzle with only 10% of the pieces. This is more pronounced in city centres where crime is even more selectively recorded (Ekblom, 1985). As a first step to understanding the problems involved in interpreting crime statistics, it is essential to acknowledge that there are three 'types' of crime, each providing a different perspective. For the purposes of this book they can be referred to as: recorded crime; actual crime; and perceived crime.

For the past twelve years the British Crime Survey has been questioning crime victims about the nature of their victimisation. The four surveys completed to date have shown that only approximately 27% of the crime that actually occurs becomes recorded in official police statistics (**Table 6.1**). There are four reasons usually given as to why victims fail to report crime, but in a general sense the decision not to report a crime is brought about by either the view that the offence was so trivial that to report it would be a waste of time

Table 6.1 Levels of recorded and unrecorded crime in England and Wales. The table is adapted from the 1988 British Crime Survey. The category of 'reported but unrecorded crime' is eliminated in order to simplify the information. In 1987, only 27% of actual crime was recorded.

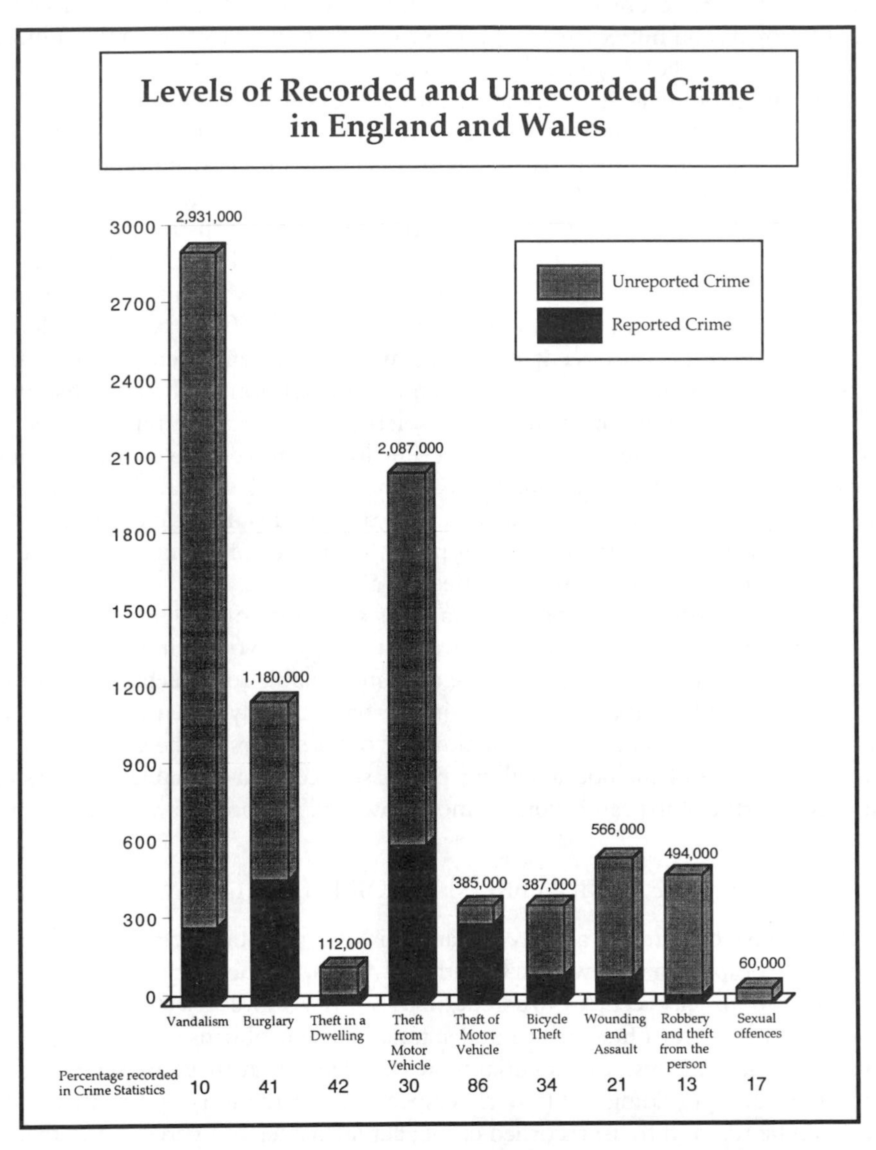

or, more commonly, that there is no reasonable likelihood of the offender ever being traced (British Crime Survey, 1980–94, and ongoing). Reported crime may not be recorded, or not recorded as reported. This can be due to factors which may include: that it is a civil, not a criminal, offence; Home Office policy; force policy; crime committed in another police area – not passed on or not recorded when received; clerical or computer error; deliberate suppression to massage statistics or miscommunication.

Accepting that reported crime may not always give a fully accurate account of the extent of actual crime, it can certainly be used as a learning tool. Each crime scene can disclose vital clues in relation to criminal opportunity, even though there is no requirement to record these disclosures in crime reports. Very recently forces are being recommended by Her Majesty's Inspectorate of Constabulary that investigators attend crime scenes together with scenes of crime officers. This collaboration, although man-power intensive, is a positive method of ensuring that the smallest piece of evidence or intelligence is not overlooked.

Whilst offender profiling is very specialised and still in its infancy, crime profiling can be applied to most crimes with beneficial effect. By constructing the profile of a crime, investigations can be assisted by more efficiently linking suspects or identifying other offences. Crimes may be analysed more accurately to produce patterns, trends, clusters and series. There are already several specialist computer software applications to aid crime analysis. Another consideration is victims who can be categorised in relation to vulnerability, naïveté, idleness and opportunity.

The 1993 Audit Commission report, *Helping with Enquiries: Tackling Crime Effectively*, suggested that Crime Pattern Analysis (CPA) was a 'key technique which can be effective against prolific criminals'. This support echoes the message that Her Majesty's Inspectorate of Constabulary has been delivering to police forces during the course of its inspection process; that police forces must be able to undertake CPA.

One of the more recent initiatives embarked on by the police is the identification and prevention of repeat victimisation. There is evidence to suggest that, in some circumstances, a building which has been burgled once is likely to be burgled again unless the factors which made it vulnerable are addressed. One study found that a burgled house was four times more likely to suffer a repeat attack than houses that had not been burgled before (Forester *et al.*, 1988). An offender knows more about the crime scene, once visited, is more comfortable because of this knowledge, is familiar with what next to steal, and how and when to steal it. Similarly the 1992 British Crime Survey found that 8% of the victims of motor vehicle theft accounted for 22% of incidents surveyed. This pattern of repeat victimisation has been shown to apply across a range of offences: racial attacks, domestic violence, crime against small businesses, crime on industrial estates, bullying, school burglary and property crime (Bridgeman & Sampson, 1994). Concerning multiple victimisation specifically against the retail sector, a 1993 survey showed that just 3% of all retailers had experienced 59% of all crimes counted (Mirrlees-Black & Ross, 1995). It also found that 2% of all retailers suffered 25% of the burglaries. These statistics give a clear indication that repeat victimisation must be taken seriously by retailers as research also shows that 24% of retailers had been burgled compared to 4% of domestic households (British Crime Survey, 1994). Precautions are now taken by the police and the Home Office to help identify and reduce factors which may lead to further repeat crimes on victims and premises.

One of the more valuable forms of intelligence gathering is by way of post-court research with offenders. Although carried out by only a few enlightened forces, this helps identify methodology and opportunity with respect to crimes committed by offenders. The aim is to successfully close down any opportunity which has been identified and reduce or eliminate the risk of further similar crimes. However, there are always dangers in closing down opportunity in one type of crime as this may deflect offenders to another, more serious, crime. Intelligence is, and always has been, one of the greatest tools in crime detection. Whilst its collection and collation are time consuming and laborious, increasingly modern technology is being provided to assist in the management, analysis and display of results in a meaningful and manageable form. Crime profiling, crime analysis and crime intelligence have a place in crime prevention. Crime analysis can aid identification of patterns of crime, evaluation of crime prevention initiatives and supporting analysis for crime prevention proposals as well as planning, monitoring and management of activities. When designing safer cities, they can each help in providing solutions by enabling planners to understand the reasons and problems which may eventually become the ingredients of crime and the opportunity for it.

Criminologists seek to learn more about the nature of criminal motivation by asking the criminals themselves (Bennett, 1984). Although this approach has ethical dangers and sometimes produces unreliable responses, it can be a means to understand the criminal psyche by exposing reasoning that can then be put to use in reducing opportunity. During police interviews of suspected offenders, the opportunity exists to uncover many secrets of criminal reasoning, but whereas the police refer frequently to criminal *modus operandi* – method of operation or 'MO' – once guilt is established or sufficient evidence exists to reasonably prove guilt, often nothing further is done to establish details of criminal reasoning. Why, for example, was a certain bank or convenience store at a particular location favoured by the criminal? Or why did a security guard or CCTV fail to deter the offender? These are considerations the police seldom pursue in questioning. Even if such reasoning were freely offered by a criminal, the police have no direct means of collating, analysing and learning from such data.

OPPORTUNITY AND RATIONAL CHOICE

There are three elements to victim crime: the 'victim', the 'offender' and the 'opportunity'. Without any one of the three elements there would be no crime. Clearly, a way to reduce victim crime is to concentrate effort on eliminating or minimising the potential for crime within one or more of the elements.

The greater the density of potential victims, such as in a crowded city centre, the greater the degree of opportunity – and consequently the more attractive the area is to potential offenders. Sociologists and criminologists argue that criminal motivation can be controlled; the urge to offend is not constant, inbred and uncontrollable but that an offender is stimulated by identifying opportunity (Bennett, 1984). Offenders exhibit rational choice in their consideration as to whether or not to commit crime. In fact, far from being a 'bumbling opportunist', the average criminal is a cunning and logical predator who

Table 6.2 Travel distances of offenders. Travel distances for three types of offence. Adapted from research by Rhodes & Conley (1981) on the distance of offences from the homes of offenders.

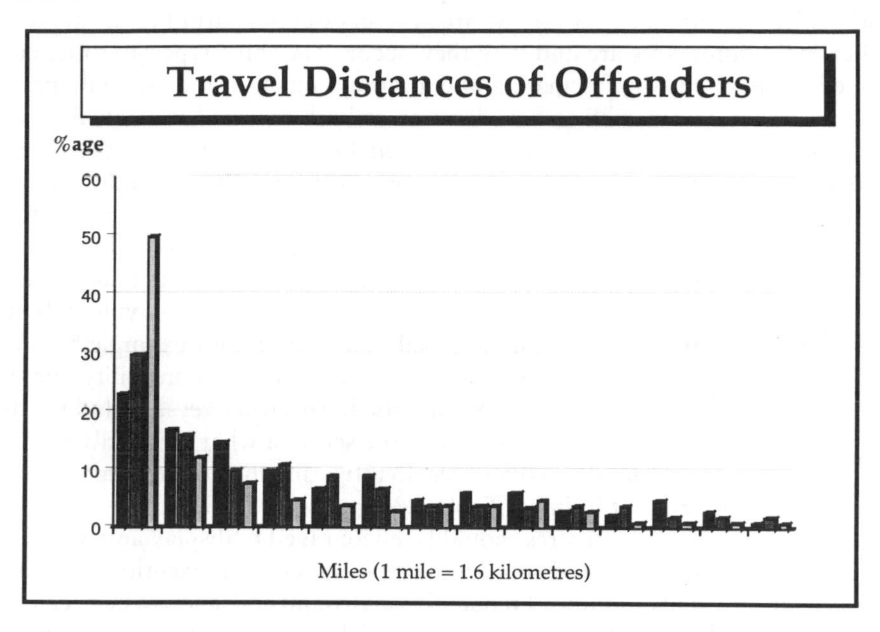

carefully selects each victim. How much longer this will remain the case is difficult to predict (Zimring & Zuchl, 1986). Drug-related crime, including those robberies and thefts committed to finance a drug habit, increasingly shows a desperate edge – the greater the degree of desperation, the lesser the degree of rational choice. Another important consideration is that, despite the increased mobility of criminals, it remains the case that most criminals need the security of local knowledge and criminal offending is usually parochial by nature – a large proportion of offending is committed within three to four miles of the offender's home (**Table 6.2**).

City centres are magnets to criminals who are attracted by the anonymity of the setting and the hope that, in a general sense, potential victims will feel disorientated because of their lack of familiarity with their surroundings and consequently feel unsure of how to respond. Sociological research conducted with offenders has disclosed a hierarchy of needs when estimating the vulnerability as a potential victim (Zimring & Zuchl, 1986). Each element that is present works to strengthen opportunity. Reasoning criminals respond to environmental cues that enable them to determine the degree of opportunity that exists. Research with criminals has shown that the value of the intended prize is a major factor in this decision making process as the offender will almost certainly not choose a difficult target unless the reward is sufficiently great. If a criminal has the choice of several potential victims from which to achieve the same successful outcome, then it is natural to select the easiest target. Retailers, who are probably the most victimised, represent easily identifiable targets for offenders, particularly in cities where an offender has plenty of choice.

PERCEIVED CRIME AND FEAR OF CRIME

As discussed earlier in this chapter, reported crime and actual crime are two quite different things; social surveys are also an essential part of understanding what those differences are and why they occur. The third type of crime, perceived crime, is equally fascinating and although much of it exists only in the imagination it is as worthy of consideration as both reported and actual crime.

The human mind works constantly to make sense of the information it processes and will use previous experiences, both directly and indirectly acquired, as a filter through which to assimilate new knowledge. The cues that we identify through the use of all our five senses are interpreted and take on a believable guise in our minds. Research indicates that one of the 'interpreters' that information passes through relates to our own sense of vulnerability (Koffka, 1935). When we feel an increased vulnerability, for example because we are carrying a large amount of cash or we have responsibility for an offspring, our levels of concern are amplified. Social surveys, including the British Crime Survey, have attempted to make sense of what is broadly termed 'the fear of crime'. A standing committee investigating the fear of crime defined a four stage spectrum of feelings (**Table 6.3**).

Ideally, visitors to city centres should be encouraged to display an awareness of crime, well-controlled and resulting in realistic crime prevention measures integrated into daily routine. Frequently environmental cues, which can be both situational or human as through physical design or through management, can skew this perception into fear, or in its more pronounced form, terror. Research shows that in females, particularly over the age of 24 years, this results in a heightened and sometimes grossly exaggerated misperception of

Table 6.3 Spectrum of feelings in relation to fear of crime (source: Home Office Standing Committee on Fear of Crime, 1980).

Fear of Crime
Spectrum of Feelings

State of Mind:	Feelings:
Complacency	Failure to take proper precautions.
Awareness, Concern	Adoption of realistic preventative measures; management of crime successfully integrated into daily life.
Worry, Fear	Preoccupation with harm and danger. Adoption of unnecessary measures to protect self which affect quality and richness of life.
Terror	Obsession. Total disruption of life.

Table 6.4 Perception of violent crime in the main shopping area of Birmingham city centre. The table compares reported, actual and perceived crime using, respectively, police statistics, extrapolations from the British Crime Survey and a social survey conducted in Birmingham in March and April 1990 (source: Poole, 1991).

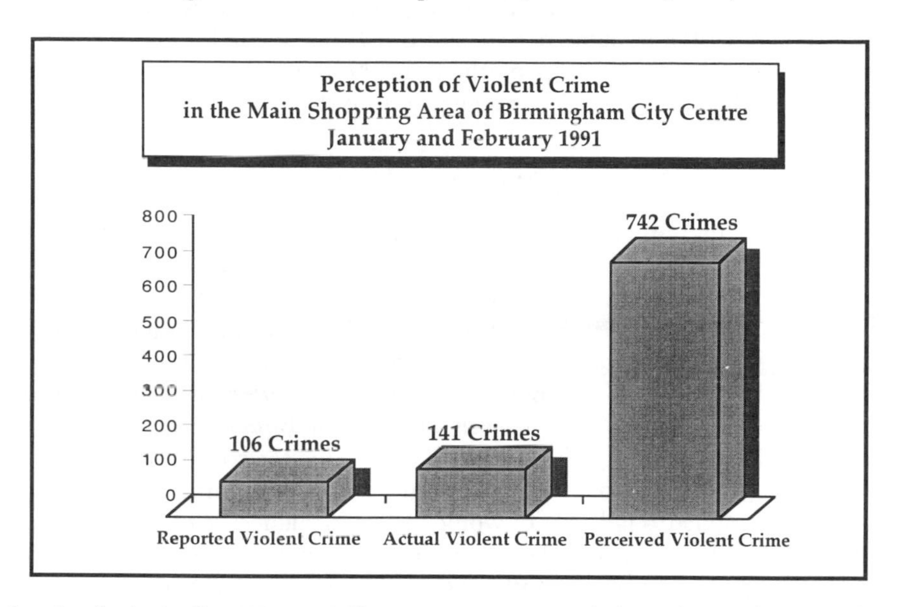

levels of criminality (Home Office, 1989). Conversely in males it often translates into a form of complacency, with the male refusing to be cowed by danger signs. The male attitude also translates into a pronounced reluctance to acknowledge weakness or to report assaults and robberies where the male is the victim. A young adult male is the most common target for violent crime in city centres.

Site-specific research conducted in Birmingham City Centre indicates that perceived crime is seven and a half times that of reality (Poole, 1991). During the survey further site-specific questions enabled the demographic mapping of geographical locations of heightened fear. When studying specific locations considered by respondents as being places where fear is heightened, it can be determined that combinations of poor design, poor management, and the presence of what American sociologists call 'people who cause anxiety' are responsible for generating fear (**Figure 6.4**). As previous surveys have shown (KPMG Peat Marwick, 1990), city centre avoidance through perceived crime can have expensive consequences not only in the diminution of vitality in city centres but also in the loss of amenities and jobs.

A heightened feeling of concern can bring on a survival mentality. Women in the Birmingham city centre surveys admitted apprehension to such a degree that 9% regularly carried a 'defence item' when in the city centre and a further 21% had identified and kept close to hand a 'comforter item' to use in times of concern.

The anonymity and loosening of social controls in cities demand overt and pronounced environmental cues that positively address fear of crime and misperceptions of the levels of criminal offending. Statistical analysis shows that an average female exhibiting acceptable awareness precautions against crime would have to visit Birmingham city centre four hours a day, six days a week for seven adult lifetimes before standing an even chance of becoming a victim of violent or personal crime, such as indecent assault or robbery. Yet to women who work or shop in city centres the perception of safety does not match the reality (Poole, 1991). In numerous social surveys on law and order, the public repeatedly ask for more 'bobbies on the beat'. Nothing, it would appear, is as reassuring as the near proximity of a guardian of the peace.

THE ANATOMY OF RETAIL CRIME

Retail crime is big business. To complement the British Crime Survey, shopkeepers now have the British Retail Consortium's Retail Crime Costs Survey. The survey, now in its third year (1996), places the costs of retail crime at £2.15 billion per annum. Included in the figure is the sum of £580 million, a 40% increase on the previous year, which retailers were obliged to spend on crime prevention. The retail price of goods is adjusted to absorb some of these crime costs to the extent that, on average, each individual in Britain spends an average of £140 extra per annum on shopping bills to cushion the cost of crime.

So impactive is crime on commerce that if, by some miracle, it could be completely curtailed retail profits would be 23% higher. Within the twelve month period of each survey there are, on average, 5.4 million crimes committed in shops – eighteen per outlet. For every ten retail premises there were six burglaries; six out of every 100 premises experienced robbery and one out of every two shops experienced vandalism. Crimes against retailers are not restricted to dishonesty. In the *Retail Crime Costs Survey*, 14,000 staff per year were subjected to physical violence and 106,000 to threats of violence. Concerning terrorism, in Britain between 1990 and 1993, no fewer than 193 bombs were placed in retail areas, resulting in, in the least serious terms, panic and loss of earnings, and in the most serious, wide-scale destruction of life and property.

Retailers, the most victimised of all groups, are notoriously low reporters of crimes, probably due to the volume that affects them. Recent research, again by the British Retail Crime Consortium in 1995 estimates that as little as 2% of all retail crime ever finds its way into police statistics. Shopping malls are an integral part of most cities, having been purpose-designed or adapted. Many thousands of people use them daily with the potential for hosts of problems. Effective crime reduction builds on people's current crime prevention activity to extend it and make it more useful. Poole (1991; 1994) has carried out research and produced two books on safer shopping.[1]

When a criminal attacks a retailer it is usually to steal items from display. The most obvious approach in defending profits is to physically increase the

1 *Safer Shopping: British and European Perspective* (Poole, 1991) examines the problems, while *Safer Shopping II: United States and Canadian Perspective* (Poole, 1994a) explores solutions and offers fifty new initiatives to help make shopping safer, control losses and enhance profits. A third book (Poole, forthcoming) will report on an experiment designed to determine the level of success of certain solutions applied in given situations.

protection around desirable items and send a clear message to a potential offender that the target has ceased to be an easy one. In shopping areas, target hardening will almost certainly displace crime. Retailers must also remain constantly vigilant to the actions of their competitors. A competitor who decreases criminal opportunity could create safer conditions in that store but may displace crime to others. A by-product of target hardening is to brutalise the environment of our city centres with hardware, such as ram-raider posts and solid roller shutter doors which can generate fear of crime. Research has shown that if offenders exhibiting rational choice are persistently thwarted in attempts to steal they will ultimately lose interest in the idea. This presupposes that a reasonable distance exists between sites of opportunity, but in the retail areas of cities where opportunities are located cheek-by-jowl the discouragement factor is reduced.

Retailers – or any potential victim for that matter – must be careful to ensure that any target hardening strategies do not encourage a criminal to commit a more serious offence. A good example of this is car-related crime, where in the last decade security has become a major issue. In the 1970s and 1980s the dramatic increase in vehicle crime, particularly the unlawful taking of vehicles for 'joy-riding', was directly connected to the inadequacy of car security systems. Car criminals often needed nothing more sophisticated than a piece of wire to succeed in the commission of the crime. As vehicle security has become more prominent, the ease with which offenders can steal unattended cars has lessened. In America the offence of 'car jacking' (hijacking a motor vehicle whilst being driven by its owner) has brought a frightening new dimension to car crime when the only way to steal a car is to steal the keys as well. What was once classed as a property crime, the theft of an unattended vehicle, now becomes a personal crime, aggravated robbery.

CRIME PREVENTION THROUGH ENVIRONMENTAL DESIGN

As there is no law requiring that a building or 'designed space', for example, a car park, has any degree of resistance to crime, the police can only advise and hope that the advice is heeded. Decisions to accept the advice will often depend upon cost factors but may be ignored altogether without any sanction. Relatively few resources are targeted towards this proactive policing activity. It is more likely that a police officer will patrol a built environment in which there is little chance of influencing change. Patrolling will hopefully be intuitive enough to result in the prevention, delaying or displacing of crime but an officer's true strength will be in responding to an incident once it has commenced or investigating an incident once it has been reported.

A positively policed city centre can attack several levels of problems simultaneously, although returning to an earlier theme, such activity and the outcomes that result will be almost impossible to measure empirically. Unless the police are prepared to accept subjective assessment in evaluating their actions, their activities will compromise and often conflict with measurable performance indicators.

POLICE INITIATIVES TO MAKE CITY CENTRES SAFER

Before outlining some of the more interesting police initiatives to make city centres safer, it is useful to summarise the limitations the police service experience in policing city centres:

- The autonomy of policing can produce a fractured and inconsistent approach to public safety.
- Crime prevention is very difficult to measure quantitatively. Police managers feel on safer ground in response-driven initiatives.
- Reported statistics give so incomplete a picture of crime that patterning and analysis are often problematic and unreliable. The identification of criminal opportunity is seldom achieved or acted upon.
- Fear of crime is as important as actual crime but the police service does not routinely analyse fear. Nor does it understand its consequences or demographically map its distribution in a site specific context.
- Despite the public wanting more 'feet on the beat', the pressures of modern policing make this increasingly difficult and because it seldom produces any tangible result it sometimes can be seen as a waste of resources.
- Only a tiny fraction of operational police resources are exclusively directed towards crime prevention, particularly at the preliminary design stage. No notice is required to be taken of any suggestions the police make to reduce or prevent crime.

Having discussed the limitations and weaknesses of the police role in city centre safety, let us now turn to the strengths and, in so doing, suggest approaches that capitalise on elements of policing that work well in cities, annotated with working examples.

Fear of Crime

One of the main indicators of safety is public perception of fear of crime, therefore consultation with local people is essential. In an effort to identify and adopt appropriate measures to reduce fear of crime the West Mercia Constabulary has carried out research and produced a package entitled *Tackling Fear of Crime – A Starter Kit* (Griffiths, 1995). It consists of a handbook on the subject, a questionnaire (which can be added to and personalised) and also computer software to assist the analysis of the resulting data. The handbook examines the phenomenon known as fear of crime and explores the direct correlation between fear and increasing crime levels. It explains the background to the questions found in the questionnaire and advises on how a survey should be conducted in order to achieve realistic and unbiased results. It also examines ways the results that such a survey might reveal can be analysed to the overall benefit of policing a particular area.

Truancy

Where are truants and what are they doing when not at school? Past examples of initiatives to return children to school include 'Sweeper' operations, where truants are collected off the streets in police 'Wagg Vans' and in more recent times, inter-agency 'Truancy Patrols' and 'Truancy Free Areas'. Evaluation of initiatives, past and present, reveal degrees of success. Truants can be deterred from entering town centres. In the West Midlands, where Sweeper operations commenced in the late 1980s, inner city crime fell by up to 30% and in Staffordshire, where the principle of Truancy Watch was established, school time juvenile arrests fell by 48%. A Staffordshire police officer has carried out research into truancy and produced a book (Lewis, 1995) entitled *Truancy – The Partnership Approach*. The book is a working document which highlights the success of partnership approaches towards problem solving. It provides valuable information to discourage juveniles from becoming victims of crime or entering areas of criminality as a consequence of their truancy.

Technology

Technology is playing an increasing part in many areas of crime prevention or crime reduction. The speed and quality by which information can be passed facilitates early notification and intervention in crimes which threaten the safety of staff or customers and the commercial success of business. 'Radio Links' is an initiative which relies not only upon technology but also on partnership. The concept is for members of radio link schemes to be equipped with personal radios to communicate with colleagues, other scheme members and the police. Differing types of schemes can be operated including retail, business and community radio links.

Two officers of the West Midlands Police have carried out research and documented their evaluation of existing schemes and identified good practice in a report titled *Radio Links – Communities Linked Together by Two-Way Radios and with the Police* (Gibson & Wright, 1995). One of the most useful ingredients of the report is a 'toolkit' to enable schemes to be created and operated in a professional and beneficial manner. It includes the important areas of constitution, protocol, procedure, agreement, codes of practice and operational guidelines all being well formulated and easily transferable. There are now a number of successful schemes in operation in many parts of the country with more planned.

Another common use of technology is Closed Circuit Television (CCTV) which can be used to good effect. There are many examples of this in private, public and partnership schemes. *CCTV – Looking Out For You* (Edwards & Tilley, 1994) is a comprehensive guide covering the background to CCTV and itemising the steps towards understanding the capabilities of the technology, planning and installation of systems. The police look forward to a time in the future when CCTV systems allow colour, tilt, zoom and maybe an audible capability as standard. Its effectiveness will then be greatly improved.

One of the more futuristic and intriguing beneficial developments of technology has been the production of 'smart water'. The concept was originated by a former police officer. The products are manufactured by the Home Office Forensic Science Service and have been endorsed by ACPO. The trade names of smart water are Probe Index and Probe Tracer. It can be a liquid, paste or gel which may be applied to any type of property or sprayed on to a criminal. It can only be seen using ultra violet light but has unique properties which allow it to be forensically examined for DNA type 'fingerprint' characteristics, thereby supplying accurate information on origin and identity. The products are already in use in a number of retail stores.

Training and Education

Training plays an important role in making cities safer. However, questions need to be asked: Who should be trained? Who will do the training? Which subjects should be included? Who will pay? These have become controversial issues and the subject of much debate, without proper direction being agreed upon or applied. The Northumbria Police, for example, joined forces with a local council to set up the country's first private security force fully trained and vetted by the police. The Derbyshire Constabulary has a scheme in operation which trains and regulates doormen or 'bouncers' at nightclubs in city centres; this scheme has been adopted by other forces.

Watch Schemes

There are many initiatives which are derived from WATCH. The public understand and support them. Shop Watch, for example, has been encouraged by the Greater Manchester Police since 1992 in the area of Altrincham, and later, Sale. It is part of a pager alert scheme. Shopkeepers may broadcast messages on scroll pagers with information on incidents such as lost or stolen cheques, shoplifters, missing or found children. Yet another 'watch' is Street Watch, which has had remarkable success. Balsall Heath is an area of Birmingham that has long had problems associated with prostitution. Residents and police of the area joined forces to make a high profile presence on the streets, during the times that prostitutes and kerb crawlers are busy, using video cameras and placards to good effect in an effort to deter the activity. Violent crime in the area has decreased by 18% and burglary by 23%.

Architectural Liaison

Architectural liaison and preventative planning are high on the agenda in the more enlightened projects that wish to ensure that building developments do not suffer from the same crime problems as others. Wisbech and Swindon are just two areas in which schemes have encouraged partnerships to develop from the planning stage through to completion and eventual occupation. The results help to ensure that cities and urban areas have a chance of staying safe from the start.

THE FUTURE OF POLICING CITY CENTRES

The police, alone, cannot make cities safer. There are no mandatory plans in being for nationwide police action to ensure that policies and strategies are set up specifically to make cities safer. However, there are many pressures upon the police to concentrate their resources far more effectively. It is far more productive to be proactive than reactive. The police do not wish to respond solely to incidents and take reports, they want to be involved in actions which help prevent or reduce crime and disorder.

Policing Plans

Policing plans, which are open to scrutiny, have to be published, as described earlier, thereby giving an open direction of local policing, which is accountable.

Police Performance

Police performance indicators are an important issue which will become even more measurable as 'league tables' emerge both locally and nationally. The need to improve performance and fine-tune the indicators will be foremost in police managers' responsibilities.

Police Resources

Financial implications play an important part of the 'pressure process'. Allocation and utilisation of human police resources are very important since they are extremely expensive with around 85% of police budgets spent on them. Budget constraints mean that financial savings have to be made in resources. Resource allocation and utilisation is one of the greatest problems for local police commanders. Cost effective deployment is paramount. Plans and initiatives to successfully combat crime that will, now or in the future, lead to fewer incidents and thereby fewer resources will provide increased productivity.

Partnerships

Partnerships are actively encouraged as never before. This will give an added dimension to planning and action as well as many other benefits, including more: resources, experience, professionalism, ownership, motivation, cost effectiveness, improved crime and criminal intelligence. Fragmented initiatives in crime prevention, crime reduction and target hardening lose impact due to their isolation, and often deflect crime to a shop, business or area nearby. In a planned partnership environment, criminals become frustrated when the opportunity to commit crime is withdrawn and will either not offend or will not visit. Safer Cities, City Pride, Crime Concern, Neighbourhood Watch, Operation Bumblebee, Crime Stoppers, Card Watch and similar centrally-funded or supported initiatives have proved that partnerships can work. We must now ensure that it can happen at all levels, locally and parochially, and for any public-dominated problem.

Technology

Technology will form the biggest thrust to make cities safer. There is an increasing development in this area, both nationally and internationally, which will go from strength to strength. Technology is becoming very specialised but surprisingly more affordable, more available and very reliable, usually providing cost efficient solutions by helping to prevent or detect crime and thereby reducing expensive human resources and improving profits or services.

CONCLUSION

The goal of policing should be to reduce crime, not simply to arrest people. The police service should identify specific problems and then target action towards those problems (Goldstein, 1990). It was Goldstein who coined the term 'problem-oriented policing' (POP). The goal should be achieved in partnership with others. Each business, organisation, service and authority, active in a city, should be aware that problems of safety cannot be resolved alone, in narrow isolation. They should take the lead, or join with others to organise partnership initiatives, co-ordinating, planning and operating relevant activities appropriate to the circumstances. It should not be only the police or other such organisations who take the first step. Funding and sponsorship are available from a variety of sources if it is necessary to purchase equipment or special services. Remember: failing to plan means planning to fail.

Our greatest weapons for creating safer cities are unity and planning at all stages of development and operation. We must be sure that we recognise the benefits and qualities already available to us – experience, knowledge, resources, stability and technology. We can all benefit from the results by saving money, making money or being able to provide a better service. But – above all – by being part of a vibrant, yet tranquil, safe environment.

7

Safer City Centres: The Role of Public Lighting

FEAR OF THE DARK

Vision is the dominant sense for all sighted humans and vision is governed by light. Light does not just enable us to see, however, it empowers us to *perceive*, to understand the three-dimensional nature of the world around us and how we interact with it. Imagine the impossibility of explaining geometry to someone who has never seen. Yet the contrast of one object against another enables us to judge distance, height and volume. So instinctive is this process that we rarely give it any thought. The capacity of eyes does not stop with enabling us to navigate ourselves through the physical environment, however, they enable us to relate to other humans. With our eyes we can sense the full range of human emotions and feelings, pain and pleasure, comfort and danger, even warmth and coldness.

Because our ability to see is so vital to life, when this ability is impaired, for example when light levels are diminished, we naturally become more fearful. Objects no longer stand out clearly one from another, shapes merge into amorphous forms, we lose our ability to differentiate colour, we cannot judge the speed something is moving towards us and perhaps most importantly, we lose the ability to differentiate friend from foe, until they are in close proximity. This last point is particularly important since our ability to recognise other humans at distance is remarkable. In broad daylight, it is only at about 35 metres that the face in effect becomes featureless. We can still however discern body movements at 135 metres; the same distance at which we can generally distinguish between a man and a woman. We can normally identify people in daylight up to 22 metres away (Moughtin, 1992). Once this distance is reduced below 15 metres, the space in which we have time to react to avoid trouble, or simply an undesirable situation, becomes reduced beyond comfortable levels.

We have become so used to living in a lighted world that our other senses, which might compensate for our incapacity to see clearly, have become dulled. Yet, for all its momentous importance, we take our ability to combat our fear of the dark, to turn darkness into light, almost for granted. In our towns and cities, we expect the street lights to turn on at night as automatically as the sun rising in the morning. In reality, however, we have a relatively unsophisticated

Copyright © 1997, Tim Townshend

understanding of the exact role lighting has in making our night-time urban environment a safer place. This chapter explores the limited knowledge in this field, based on research over the last two decades, reviews preliminary research that has been carried out in Newcastle examining the role of lighting in city centres and finally suggests that much more vital research has yet to be carried out.

ILLUMINATING THE STREET

The last time people in British towns and cities had to think carefully, for a prolonged period, about going out after dark was during the 1939–45 War. In an anachronistic reversion of modern experience, with the absence of street lighting and no light from windows, people were returned to the conditions of medieval times (O'Dea, 1958). Even with the perils of war-time, however, the streets were in many ways safer than before the introduction of street lighting. During the Middle Ages, not only was venturing out after dark hazardous, but it was a necessity to carry a lantern, as not to do so was liable to be regarded as suspicious and might well lead to arrest (Salusbury-Jones, 1975). Carrying a light was as much to make yourself visible to others as to illuminate the way for yourself. There was an acceptance that every light represented another human being. There existed a balance between individuals: observe and be observed; although naturally it was an unstable balance, as an individual could choose to alter it by extinguishing their own light.

At this time, lighting was seen as the only effective means of controlling anti-social nocturnal behaviour. The monopolisation of street lighting through state control and supply, therefore, promised security and was generally welcomed by the public with its introduction in Europe from the late seventeenth century. Under certain regimes lighting was under direct control of the police. Pre-Revolutionary Paris is a good example of where lighting was introduced by order of the King to protect respectable citizens from 'commotion and uproar' (Gottingen, 1756). Street lights were deliberately hung on wires in the middle of the street and although Parisians criticised their interference with traffic, they were a purposeful display of who lit the streets and who ruled them. These lights were dim and so did not illuminate anyone particularly well: the psychological impact, however, was clear. The implicit link between light and observation during this period implied that people were watched by the police at night. In reaction to this, lantern smashing was a fairly common occurrence in pre-Revolutionary Paris, as destroying streets lamps had the 'satisfaction of unseating the authority they represented' (Schivelbusch, 1988).

From these early beginnings, the supposed link between crime prevention and lighting became strong. Even as late as the 1920s, lighting was being sold in the US as 'a policeman every 50 yards' (Holden, 1992). It is interesting to note, that while there are obvious differences in the two technologies, the parallels between the aims and objectives of the introduction of early street lighting and the current installation of CCTV in many of our towns and cities is remarkable. Even the language and metaphors, particularly, for example, in the depersonalisation of policing, is similar. CCTV is discussed more fully in Chapter 8.

LIGHTING AND CRIME

The proposition that night crime could be impacted by the introduction of street lighting, therefore, is based on the premise that good lighting increases the ability of respectable citizens and the police to observe miscreants, thereby acting as a deterrent to criminal activities. Though this does depend on crime being a covert activity where the risk of being observed is unwanted, which may not always be true, this intuitive approach received great impetus through the work of Oscar Newman (1972). The principal concept here is that the moderator of offender behaviour is social control brought about by surveillance and the observation of others.

In addition to light enabling people to see and be seen, is the fact that since people are more fearful after dark, reducing the impact of darkness will in turn make people less afraid to venture out. This increases the number of people using public spaces at night, increasing chances for observation and social control, reinforcing the preventative nature of lighting and, in turn, returning quality to night time public life. It was also supposed that the aesthetic properties of good public lighting can make urban areas attractive places to visit in their own right, again adding to urban quality. Experience has shown that these cause and effect relationships may, however, be far from proven.

Between December 1973 and February 1974, British streets were blacked out, albeit temporarily, as a result of emergency power measures. The impact of this event on crime rates did not go unnoticed. Brighton and Sussex Police, for example, reported that burglary increased by 100% while thefts from vehicles rose by 59% during this period (Mayhew, 1976). Though this trend was by no means universal – Newcastle-upon-Tyne, for example, showed no increase at all – the possibility of a link between street lighting and crime rates could not be overlooked. Moreover, concurrently a number of studies were emerging from the US which were also exploring lighting, crime and fear.

Unfortunately these studies produced widely varying results and conflicting evidence. Jones (1975), studying a re-lighting scheme in New Orleans, concluded that new high intensity lighting had negligible effect and that patterns of offence were not appreciably different before and after the scheme. In the same year, a District of Columbia study showed a 30% reduction in crime after a re-lighting scheme, with further reductions with subsequent additions to the project (Hartly, 1975). These two studies have been widely used to illustrate the contradictory nature of lighting and crime prevention evidence.

So contradictory were the results that the US Department of Justice commissioned a review of 103 separate schemes nation-wide. The findings of this too, however, were inconclusive. The report claimed that nearly all the studies were based on inadequate, or partial, understanding of the effects of different types of light and most had employed flawed analytical techniques. One subsequently positive and important point did emerge from this study. This was that while effects on crime rates might not be conclusive, there was a strong indication that the fear of crime was generally reduced by improvements to poor lighting (Tien *et al.*, 1979).

Unfortunately most of the US research and subsequent UK work has centred around residential areas, rather than mixed used, or commercial, city centres. Great caution must, therefore, be employed when considering their results in the light of a quite different cultural arena. There is, however, a great deal to be learnt from these studies and they highlight a number of topics that might usefully be examined in a city centre context.

LIGHTING, CRIME AND THE FEAR OF CRIME

The realisation that crime and that the fear of crime are two separate issues became important from the early 1980s. The likelihood of someone falling victim to a crime and their fear of that crime happening not only appeared unrelated, but whilst crime rates were rising, people's fear of crime was rising at a much faster rate (Hough & Mayhew, 1983). The issues here are complex, but often the least likely victims of recorded crime, elderly women, for example, are understandably the most fearful. The issues here are particularly complex, since those members of society shown by British Crime Surveys as the least likely victims of crime may be so simply because they impose curfews on themselves, i.e., they are simply not around to be attacked. The most influential work to relate this directly to lighting, at this time, was carried out by Painter (1988; 1989). Her evidence suggested both crime and fear of crime could be reduced by improving lighting at particular trouble spots, in relation to threatening behaviour, vandalism, personal attack and auto-crime and, in particular, was beneficial to the lives of those most fearful: women and elderly persons.

Painter's work is perhaps most pertinent to this book because, whilst not in a city centre, it was set in inner city areas which were not purely residential. Her work has, however, subsequently attracted some criticism, not least because the work was carried out over quite short time scales. Other research has shown that improvements in crime rates brought about by physical interventions may taper off after a period of months (Allat, 1984; Ramsay, 1989). Further criticism surrounded the fact that studies looked at small scale localised black-spots, such as walkways and tunnels, where re-lighting might be most beneficial. Although small scale projects can create methodological difficulties, due to the reliance on small sample sizes, this does not necessarily detract from the work. Some of the subsequent larger scale projects have recognised their own problems, such as isolating and evaluating the change of a single environmental factor, such as lighting, over a large area (Herbert & Davidson, 1994). The work also did not include any search for crime displaced by the improvements in her study areas. Displacement has been discussed in Chapter 3. In relation to lighting studies, however, evidence is very limited, although an early US project suggested crime displacement of around 25% to neighbouring areas (Wright *et al.* 1974).

Finally Painter was criticised because her work 'under concentrated' on actual crime and 'overemphasised' harassment (Ditton *et al.*, 1993). The last criticism may be rather misjudged. Increasingly feminist writers have stressed the impact of harassment on the lives of women, such as being kerb crawled, or called after in a sexually aggressive manner, which are not treated as crimes by the police and criminal justice system, yet create feelings of victimisation. It is quite possible for these sub-criminal activities to have a devastating effect on victim's lives (see for example, Pain, 1991). Racial abuse may have a similar effect (Commission for Racial Equality, 1987). The emphasis on harassment by Painter's work should then be seen as a strength rather than a weakness, as should the clear separation of actual incidence of crime from the fear of victimisation. The fact that the work showed significant reductions in both sexual and racial harassment should be seen as significantly improving the quality of life of those affected.

BRITISH PARLIAMENTARY LIGHTING GROUP PROJECT

By the 1990s, therefore, research was still disputed, though as Atkins *et al.* (1991) point out, given the many factors which influence perceived risks, contradictory results are 'not surprising'. The result was a five city survey commissioned by the British Parliamentary Lighting Group Project in 1990/91. The areas studied under this project all had certain characteristics in common. The overriding factor was that the areas were exclusively residential; The Dukerics, Hull, studied by Herbert & Davidson (1994), for example, was an area of 1890s terraced housing. In addition, areas needed to exhibit a range of social problems, although they were not perceived as problem areas *per se*. While generally they displayed poor lighting standards, they did not have particularly high crime rates. The methodologies employed were also relatively standard for the areas, with household surveys carried out before and after re-lighting, backed up with secondary data. Although applying the findings from these studies directly to city centre locations would be dangerous, it is useful to examine their key findings.

The general finding was that people's attitudes to improved lighting standards were positive and they felt that their areas were safer at night. There were, however, a number of inconsistencies and the results were perhaps less unequivocal than might have been wished for. The Strathclyde team, for example, reported that pedestrians had increased feelings of security when out at night, with greater numbers prepared to walk alone after dark. This was backed up by observations made of the number and demeanour of, in particular, women and elderly persons out on the streets after dark. When respondents were asked about their fears of falling victim to specific crimes, however, these did not diminish significantly after re-lighting compared with the before re-lighting survey (Ditton, 1993)

Several of the case studies highlighted a gender divide over attitudes towards improved lighting. In particular, there were reductions in fear after lighting improvement schemes as perceived by women themselves (Herbert & Davidson, 1994) reinforcing Painter's earlier work (1989), or more, as perceived by the community as a whole (Atkins *et al.*, 1991).

In Hull, there were large reductions in the feeling that it was unsafe to go out after dark and people were less worried about youths hanging about. Some aspects of drunken behaviour, however, such as shouting and fighting between drunks, became more noticed after re-lighting. Herbert & Davidson (1994) suggest this is a dual process of 'visibility and exposure'. The fact that the new lighting made them more visible, compounded by the increased numbers of residents out on the streets after dark, made people more aware of the problem.

The realisation of the potential paradox that improving street lighting may make some crimes more visible is important. So too, however, is the acceptance that some crimes will occur regardless of lighting levels. This may be due to factors outweighing any perceived disadvantage of being seen to the criminal, or simply because lighting levels are irrelevant, for example, drunken brawls outside pubs which may be well lit. Studies of offenders suggest that lighting and visibility is not always a prime consideration for the criminal (Ramsey & Newton, 1991). Moreover around a third of recorded crimes are committed during daylight hours (British Crime Surveys). These complexities do not detract from the value of studying the effects of lighting on crime, or the promotion of better lighting in city centres, they merely demonstrate that generalisations about behaviour are extremely dangerous.

LIGHTING IN CITY CENTRES: THE NEWCASTLE SURVEY

Given that the overall finding of the Lighting Group Project was that people react positively to improvements in lighting, can this lesson be applied to city centres? The greatest problem here, of course, is that most city centres are already lit to high standards (at least to BS5489, the British Standard for road lighting) and certainly better than the residential areas that made up the early 1990s study. It might be questioned, therefore, whether lighting could actually be improved in many city centre streets without excessive cost and wasted energy. On the other hand, it is the city centre where people will most likely encounter the unfamiliar and, for example, wish to ask for directions. Doing this after dark may well be more intimidating, especially, for example, for a single woman.

The key issue here, however, is that better lighting does not necessarily mean brighter lighting. Cities' centres may on the surface seem brightly lit, but this does not mean they are well-lit in terms of people's safety, or in addressing their fear of the dark. Much city centre lighting is still aimed primarily at the safety of the motorist rather than the pedestrian and often fails to address adequately pedestrian needs. Issues such as an even spread of light, good colour rendering and elimination of dark shadows may be as important for the pedestrian as the levels of luminance. The city centre is also the most likely context for public lighting as display, for example, floodlighting buildings. We need to understand what kind of impact this has on people's experience of the city at night.

The preliminary city centre survey in Newcastle was aimed at examining the main issues outlined above, as well as attempting to explore some of the other identified topics, such as lighting and different gender responses. The work was carried out over two popular nights out, Thursday and Friday, during March in the early evening, in a well-known pedestrian area around Grey's Monument, a popular meeting place in Newcastle's city centre. The city centre of Newcastle is generally lit with low pressure sodium (orange) lighting. In addition, some areas have heritage reproduction lighting and floodlighting of buildings.

Given that the most positive aspect that had come through previous research was the relation between lighting and fear reduction, the survey was aimed at people's perceptions of crime, safety and feelings of well-being. The respondents were asked a series of questions relating to: lighting in the city centre and their feeling of safety, lighting and perceived effects on crime, areas in the city that they felt were badly lit, comparisons with CCTV and feeling of safety, plus their opinions concerning the flood-lighting of buildings. There was no attempt to compare their perceptions with actual reported crime levels in the city centre.

The majority of respondents to the Newcastle pilot study felt that good street lighting helped cut some crime and, more specifically that levels of personal attack/assault and car theft/break-in would be lower in well lit areas. Seventy five per cent of people stated that good street lighting made them feel safe at night, while people who thought that the city was badly lit during the evening were twice as likely to feel unsafe in the city after dark than those who considered the city well lit. Furthermore, two thirds of respondents who felt good street lighting was not a particularly good deterrent against crime also stated that good street lighting made them feel safer. This last point strongly suggests that even when people recognise that crime may not be affected by lighting, the psychological impact of light, making individuals feel safer, is of greater importance.

This would appear to be in line with previous research. Previous studies, however, have all been associated with lighting improvement schemes and therefore potentially open to criticism because any perceived benefits may be short lived without any real gain in quality of life. The work in Newcastle would tend to dispute this, as it was not carried out with reference to improvement schemes and yet still indicated a strong link between good lighting and levels of personal well-being.

Relative Light Versus Absolute Light

Early work from the Massachusetts Institute of Technology suggested that pedestrians were less reassured by overly bright lighting schemes than by those which threw an even spread of light over an area (Hack, 1974). In the Cardiff and Hull studies, respondents thought the re-lit areas were much brighter and better maintained. In fact according to Herbert & Davidson (1994) much of what was achieved was a better distribution and a 'sharp reduction in dark areas'. The work in Newcastle attempted to explore the relationships between actual levels of street light and perceived levels in relation to people's sense of security, by comparing areas of the city people thought were well lit with those they thought were badly lit.

There was a great deal of consistency in people's responses to questions on lighting levels. Two areas were strongly identified as dimly lit and unsafe: Manors just to the east of the city centre and the Dean St/Quayside area of the city. Of particular interest was the continued reference to Dean St/Quayside. The Quayside area of riverside has been greatly regenerated over the past decade and, in fact, is well lit. Dean Street is the main access to the Quayside. To reach both, however, the most direct route from the city centre takes the pedestrian along Grey Street. During the early 1990s Grey Street underwent a re-lighting programme, with the erection of reproduction nineteenth century columns. However, these were felt to give too little light so that high level flood lighting was introduced. The result is that high levels of light wash over the buildings giving the whole street a vibrant orange glow. Though it is far from proven since none of the respondents made the connection between the brightness of Grey Street and the relative levels of lighting in Dean Street, it is quite understandable for members of the public to see Dean Street – a continuation of Grey Street and where the high level lighting comes to an abrupt end – as relatively dark in comparison.

There are also a number of poorly lit pedestrian alleyways and flights of steps which lead from the quayside to the higher level of the main central area. Again whilst most pedestrians would not use these after dark, it was felt by respondents that these were potential areas for attackers to hide. Thus, even if the main thoroughfare is brightly lit, the propinquity of areas in dark shadow can cause concern for the pedestrian.

Lighting and Gender

An unexpected aspect of the Newcastle study was the relatively high levels of insecurity expressed by male respondents. In the 16–25 year old age group who made up over 50% of the survey, just under a half of male respondents did not feel safe or felt very unsafe in the city after dark. This was lower than their female counterparts, two thirds of whom felt unsafe, but was still unexpectedly high, especially given that they were being asked about the centre generally, rather than about particular trouble spots. Painter (1989), for example, found levels of fear of 51% and 73% for males and females respectively; this, however, was in relation to a particularly bad location. Moreover, of those males who were fearful, more than half responded positively when asked if good street lighting would make them feel safe. This was only slightly less than for females, 60% of whom reacted positively to the same proposition.

Research has suggested that men are less likely than women to express their concerns regarding safety (Vamplew, 1990; Maxfield, 1984). The fact that the Newcastle survey showed a surprisingly ready admittance to fear by men and the appreciation of lighting may be due to several factors. It is strongly suggested, however, that of real significance is location. As previous lighting studies have generally been located in residential areas, they will reflect how men feel about the area in which they live. There may well be a reluctance to admit to fear on 'home ground', particularly in a head of household survey. In the relative anonymity of the city centre, this reluctance may – to an extent – disappear.

The observation that the fearful men reacted positively to the suggestion of good street lighting may also be a factor which is different in the city centre than in a residential area. The need to identify trouble, for example, rival football supporters, may be more acute in the centre than in a residential area where the men might be less likely to meet a stranger. This is not to contradict existing research, since the fears of men were less acute than those of women: clearly the intimidation and vulnerable feelings of women must take primary importance. It does, however suggest that the fears of men, particularly young men, in the city at night may also need serious consideration. More importantly, if there are these large numbers of young men harbouring concerns over safety, it may influence their own behaviour in their reaction to fear, outward aggression and displays of machismo, increasing intimidation of others. Although there may be little that good lighting can directly do to address this problem, it is undoubtedly a problem that must be addressed by all working towards safer city centres.

Lighting and CCTV

There was little gender difference in the attitudes to the relative merits of lighting and closed circuit television (CCTV). Although examining people's reaction to CCTV is a quite separate topic, it was useful to ask comparative questions since much of Newcastle's city centre is covered by CCTV cameras. Previous research has also suggested that in limited situations – for example, subways – lighting may be more influential for safety than CCTV (Atkins *et al.*, 1990). In terms of people's perceptions towards crime, however, the respondents were almost unanimously agreed that CCTV was a better deterrent against criminals than good quality lighting. There was, however, an even split when asked which technology made people feel safer. Although the reasons for this were not always made explicit, it was clear, for example, that some respondents mistrusted the technology of CCTV. One respondent summed this up succinctly: 'you can see when an area is clearly lit, but how do you know if a camera is working even if it is pointing at you'. One respondent talked about CCTV giving a 'false sense of security', whilst others had witnessed either criminal or violent events in areas covered by CCTV and felt the cameras did little to deter trouble. Again as with lighting, those involved may not care about being observed if, for example, their judgement is impaired by drink.

Research has suggested that certain groups in society may feel that they are targeted by CCTV and, therefore, distrust its installation (Mihil, 1993). There was, however, no discernible pattern from the Newcastle research, apart from the observation that people did not necessarily feel reassured by its presence. Overall, therefore, CCTV may almost have a reverse profile to lighting in that it may deter crime, but does not always instil feelings of personal safety. By contrast, lighting may do less to reduce crime, but may be personally reassuring.

Colour Rendering

The colour of lighting was not specifically covered by the Newcastle work. The topic was however raised by several respondents, because residential areas of the West End of Newcastle have recently benefited from new white (high pressure sodium) lighting, funded through City Challenge. Those who mentioned the new lighting preferred the white lights, in comparison with the city centre's orange (low pressure sodium) ones. The ubiquitous orange/yellow glow of the typical British street light is generally favoured by highway engineers due to their cost and efficiency. Designed for highway use where mostly the colour of other traffic is generally unimportant, the colour rendering[1] ability, however, is virtually non-existent.

It is interesting to note that the Strathclyde study, undertaken as part of the 1990/91 parliamentary study, attempted to test whether better colour rendering had a greater effect on crime or fear of crime by re-lighting one site, in Glasgow, with high pressure white lights and the other study site with orange low pressure sodium lights. Overall the study concluded that in terms of recorded crime there was no superiority of one system over the other. In Bellgrove (white lights) police recorded crime fell by 46%, whereas in High Blantyre (orange lights) it fell by 43% (Ditton *et al.*, 1995). Interestingly, however, although a higher percentage of residents in the orange lit area were prepared to walk alone after dark, they actually felt less confident than those in the white lit area. Furthermore when pedestrians were asked about their preferences, the majority in both areas favoured white lights.

There is a strong suggestion, therefore, that good colour rendering may increase people's sense of well-being, even if actual crime/incident rates are unaffected. Again this needs to be looked at more carefully with respect to the city centre where contact with strangers is more frequent and objects, buildings, etc., are less familiar. Anything that can reduce the fear of the unfamiliar needs serious consideration.

Floodlighting and Lighting for Display

When examining the successes and failure in lighting in British cities, the Royal Fine Art Commission (1993) found more to be critical about than to praise, particularly with regard to the inappropriate floodlighting of buildings. There would appear to be a wasted opportunity here, as the floodlighting of buildings is potentially popular. The successful floodlighting of individual buildings, particularly when part of a co-ordinated scheme for the city centre, can do much to make the night time urban environment more attractive and thereby encourage the use of public spaces after dark. Eighty-five per cent of respondents in the Newcastle survey felt that the floodlighting of major buildings was either important or very important to the night time image of the city. Respondents felt that it was particularly important for structures such as the Tyne Bridge, which are sources of civic pride and identity, to be lit up. It is also interesting that none of the respondents mentioned the floodlighting of Grey Street which suggests that it has not succeeded in attracting people's attention to the street – although some people did mention a particularly prominent building within the scheme, the Theatre Royal. It is possible, therefore, that excessive floodlighting not only potentially gives adjacent streets a negative image as already discussed, but also, by reducing the contrast, lessens its impact on the viewer.

In relation to the lighting of buildings, several of the Newcastle respondents mentioned shop and shop window displays. More research is needed here, but it is likely that they play an important part in the image of the city at night. This has direct implication for issues of shop security measures and would support the resistance to solid shutters being fitted to shops.

1 Rendering is the colour an object is perceived to be when compared with daylight. Some forms of lighting have particular good colour rendering, others are extremely poor (see, for example, Chartered Institution of Building Service Engineers, 1992).

Lighting Pollution

No review of lighting in our cities would be complete without considering the topic of light pollution. First raised as an issue in the US, where the scattering of artificial light by dust particles and water droplets was obscuring the view of astronomers, the issue has broadened to encompass concerns about energy efficiency (Institution of Lighting Engineers, 1992). It seems particularly pertinent in relation to discussions of safety too, since badly designed security lighting can be a specific problem (Royal Fine Art Commission, 1993, p.13). Although the Newcastle survey made no mention of light pollution, some respondents raised the issue unprompted. This might, in fact, suggest that the term light pollution, in the same way as other environmental issues, has entered common parlance and is likely to become an increasingly public concern.

CONCLUSION

The main conclusion to this chapter must unfortunately be that at present knowledge of people's perceptions of and attitudes towards public lighting after dark is, at best, partial. Most lighting research has centred around residential areas and has been specifically aimed at assessing the changes in levels and fear of crime brought about by specific re-lighting programmes. This has created a knowledge base, but one that is specific to residential areas.

Most of the research in residential areas has suggested that there is no link between improvements in lighting and reported crime rates over wide areas, although small scale targeting of particular areas may be successful. More important, however, is that the general public still display a great, though not limitless, faith in lighting – if not to deter crime, then, at least to make them feel safer after dark. Not only this, but they continue to attach importance to good lighting despite the introduction of new technology such as CCTV.

In studying the effects of good street lighting it should not be ignored that, as with all physical interventions in the environment, the root social causes of crime are not being addressed, although people's fear may be. Fear is a very destructive force. If people cannot feel safe in their city centre, then it is unlikely that they will be able to enjoy a night out at the theatre, cinema, pub, or other venue. Gradually people will abandon the city centre for the relative security of the heavily guarded out-of-town complex, or even the safety of home entertainment. The city centre becomes abandoned, urban life with all its culture and vitality dies.

No one would suggest that good lighting can stop urban decline, but if good lighting is as important to an individual's sense of well-being as research suggests, then city authorities cannot afford to ignore its impact. Few British city authorities have lighting policies in their statutory plans, planners often display little knowledge of the technology or, perhaps more importantly, of public perceptions to lighting, while urban designers often get too involved in issues of aesthetics. In the future, it will be important to ensure that the aspirations of ordinary members of the public are reflected in the management of our cities after dark, and that appropriate lighting enables the reclamation of a truly democratic, night-time public arena.

8

Safer City Centres: The Role of Closed Circuit Television

Almost imperceptibly, over the last ten years, closed circuit television cameras (CCTV) and systems have become a pervasive and common feature of city centres, car parks and residential areas – as much a part of the British landscape as telephone booths and postboxes (Campbell, 1995). The first public CCTV system in Britain, consisting of eighteen cameras, was installed in 1985 in an attempt to counter vandalism along Bournemouth's sea front. A more extensive installation also began in 1985 when – after a particularly bad year of football hooliganism – the Football Trust charity offered grants to every professional football club to help establish CCTV systems in their grounds (Davies, 1996, p.186). Including private systems, there are over 150,000 professionally-installed CCTV cameras in British towns and cities; over five hundred more are being installed each week (Graham *et al.*, 1996, p.2); more than eighty local authorities have installed some form of CCTV system in their town or city centre, while more than two hundred CCTV schemes in public places have been started. Cameras are even being considered for small rural towns.

Dramatic reductions in the incidence of crime are used to demonstrate the effectiveness of CCTV. For example, figures from Airdrie show that, in the first year of operation, crime fell by 75%; car crime was reduced by 94% as were commercial break-ins; vandalism fell by 84%; public order offences were cut by 42%; and the number of serious assaults was halved (Scottish Office, 1994, p.30). In addition, many sceptics were persuaded of their value by the high profile role played by CCTV in early 1993 in the police investigations of a Liverpool murder case and terrorist bombings in London (Geake, 1993a, p.19). In mid-February 1993, television news programmes broadcast grainy images of a two-year-old boy apparently being led away by two youths. The boy, Jamie Bulger, was later found murdered. In March 1993, television again broadcast pictures of two men recorded on security cameras at Harrods and Victoria Station shortly before bombs went off. In each case, arrests soon followed: for the bombers, within hours of the television broadcast; for the murder of the child, within days. A pernicious and potentially dangerous assumption, however, as Graham *et al.* (1996, p.3) warn, is that technology of this type can provide some 'quick technical fix' to the complex social problems surrounding crime and police safety.

CLOSED CIRCUIT TELEVISION SYSTEMS

The installation of CCTV systems in many British cities has generally been seen to be a 'good thing', forming an apparently essential part of many city centre revitalisation projects and the subject of regular eulogy by Home Office ministers (for example see Home Office 1995b; 1996a; 1996b; 1996c; 1996e; 1996f; 1996h). One Home Office Minister has called them the 'friendly eye in the sky', arguing that 'there is nothing sinister about it and the innocent have nothing to fear. It will put criminals on the run and evidence will be clear to see' (from Campbell, 1995). To further encourage their installation, the British Government removed the need for planning permission for camera installations[1] and has provided funding for systems through a competitive CCTV Challenge programme for local authorities and other agencies who also have to provide matching funding. Two rounds of the challenge have been held. The first in October 1994 offered £5 million and attracted 480 bidders with 100 winners; the second in November 1995, offered £15 million and attracted over 800 bidders with 250 winners. A third challenge, again offering £15 million, was announced in August 1996 with the Government's intention being to put 10,000 more cameras on the streets over the following three years. As an additional incentive in some cities, insurance companies – whose business is ostensibly based on fine judgements about risk – offer discounts on premiums to retailers contributing towards the costs of CCTV systems.

Few venture to criticise CCTV. Graham *et al.* (1996, p.4) note the criticism that does emerge is usually deflected by 'the claim of technological neutrality which implies that any critic . . . [is] somehow pro-crime'. Further emphasising the apparently benign role of CCTV cameras, the ACPO Working Party on CCTV (from Honess & Charman, 1992) argued that: 'The camera, recording a public place, sees no more nor less than a plainclothes officer sees and is certainly a more accurate record'. The general police response has been that rather than restricting public freedom: 'CCTV surveillance *increases* public freedom, enhancing opportunities for people to enjoy public places' (Arlidge, 1994, from Graham *et al.*, 1996, p.4). The concern for public safety therefore seems to outweigh any potential infringement of civil liberties.

Research by the Home Office (Honess & Charman, 1992) suggests that, in general, citizens seem to have few concerns about the invasion of privacy and potential erosion of civil liberties aspects of these systems. Graham *et al.* (1996, p.4–5) note that a survey in Glasgow, where a 32 camera system became operational in November 1992, showed that 90% of people supported the project, 66% believed the system would make the city centre a better place and 40% said it would make them visit the city centre more regularly. Nevertheless, aggregated statistics may mask a more complex picture. Young men are generally more suspicious of the systems, especially young black men who already feel excluded from shopping malls where they experience intense scrutiny from security guards (Evans, 1995, from Graham *et al.*, 1996, p.5). Young

1 Davies (1996, p.187) notes that removing the need for planning permission is also a means of stopping local authorities blocking the installation of camera systems.

women, who may support CCTV in reducing petty crime, did not feel that the threat of attack was reduced when compared with other measures, such as improved street lighting or increased police patrols (Honess & Charman, 1992, p.11). In researching this book, the authors heard from women accounts of being assaulted while ostensibly being observed by CCTV cameras.

SURVEILLANCE

As discussed in Chapter 3, the deterrence effect of surveillance relies initially upon the assumption that, due to increased visibility and the risk of being seen, the would-be offender is discouraged from deviant behaviour. Where the offender is undeterred by merely being seen in the act, surveillance will only operate effectively where there is the strong possibility of intervention by what Cohen & Felson (1979) term a 'capable guardian'. CCTV systems increase both the visibility aspect of surveillance and – at least in principle – the expectation of some form of intervention by a capable guardian should an offence be observed. In the latter instance the response time is clearly important in terms of averting the offence. Although the panning of the camera over its survey area may mean it misses some incidents, one of the other advantages of CCTV is that – at least on tape – it provides 24 hour coverage. CCTV systems need to be evaluated in terms of both their deterrence effect and their effect in terms of the detection, apprehension and prosecution of offenders. The effect of cameras on people's fear of crime must also be considered. Geake (1993) reports a security consultant stating that 'the effect of CCTV is 95% deterrent, 5% detection'. As Graham *et al.* (1996, p.2) note this has led to dummy cameras being installed. Incidents that occur in front of dummy cameras can, however, undermine public confidence in the whole system.

THE EXPERIENCE OF CCTV SYSTEMS

In his research report, *CCTV in Town Centres*, for the Home Office Police Research Group, Brown (1995, p8) argues that: 'CCTV cameras can help the police to tackle crime and disorder by improving capable guardianship and increasing the risks associated with offending. This increase in risk reduces the suitability of the target and de-motivates the offender.' More specifically Brown (1995, p7–8) notes that the cameras increase capable guardianship by acting as:

- **an aid to deployment** – camera operators can 'patrol' city centres effectively and efficiently, and because of the carefully selected siting of the cameras, will gain an excellent view of incidents as they start to occur
- **an aid to the identification and arrest of suspects**
- **a deterrent to criminal/offensive behaviour** – the very presence of cameras and the publicity generated by schemes may act as a deterrence for offenders
- **an evidence gathering tool** – the cameras are not only used to film incidents as they occur but are also used to film the police response.

Brown (1995, p.8–9) also notes that the effect of cameras on different types of crime may depend on two factors. First, the nature of the area under surveillance (large and complex versus small and simple layout). Secondly, the nature of the offence, i.e., whether or not the crime is committed surreptitiously, and the extent to which an offence is either impulsive and affective (as with rowdy behaviour) or instrumental and planned as, for example, with robbery of a bank. The choice perspective of crime suggests that CCTV should have a greater impact on instrumental or acquisitive crime and less on affective or expressive crime.

CCTV in Newcastle and Birmingham City Centres

Brown's (1995) research looked at the impact of CCTV systems in a number of cities. The following sections summarise and review his research in two cities: Newcastle and Birmingham.

Newcastle-upon-Tyne

Newcastle-upon-Tyne is a large provincial city in the north east of England. Its town centre is compact but, in many ways, is typical of most English metropolitan cities. The centre has a low residential population but a large number of pubs, night clubs, restaurants, shops and offices. In December 1992, a 16 camera monochrome CCTV system was installed with camera positions selected using crime pattern analysis (CPA). Initial funding for the system came from the City Centre Partnership Security Initiative – a corporate initiative set up using a grant from the Department of the Environment – and funds from the local private sector. The local police authority is responsible for the ongoing maintenance cost plus the salaries and costs of the system's civilian operators.

The area covered by the cameras contains a number of major vehicular thoroughfares. It is partly pedestrianised and made up of shops, commercial and financial properties, with an extremely high number of licensed premises. The Eldon Square covered shopping mall within Newcastle city centre also has its own privately operated camera system. All of the cameras have a pan, tilt and zoom function. The city centre is conducive to camera surveillance: its streets are wide and relatively straight; and there are few subways and few obstacles which block the view of cameras. The area covered by each camera is considerable and overlaps with those areas covered by neighbouring cameras. Very few streets in the centre do not have some form of camera coverage, while premises which had previously been identified as the most vulnerable have full camera coverage.

Birmingham

Birmingham is the UK's second largest city. Like Newcastle, its centre has a low residential population, but has many shops (including three large and one small shopping centres), offices, and licensed premises. Extending over a larger area than Newcastle's, the city centre has a more complex layout and a greater number of natural obstacles to coverage by cameras. Furthermore, at the time of Brown's survey there were only 14 cameras – two fewer than in Newcastle city centre.

The popularity of Birmingham city centre increased during the late 1980s as Birmingham became a popular venue for rallies, demonstrations, protests and marches for a wide range of organisations, with general public safety within the city centre becoming a major issue. In 1989, the local police commander suggested to the Birmingham City Centre Association that CCTV be installed within the city centre. The intention of the system was not only to help police large demonstrations but also to make Birmingham a safer place through tackling the problems of general street crime such as: robbery, theft from the person, criminal damage and assault. The association approved the suggestion and, in 1989, the Citywatch Trust was formed.

Citywatch proposed an ambitious four phase plan to install 27 cameras together with the infrastructure for an additional 21 cameras. The first phase involved the installation of cameras in the central core of the city; and in the second phase, the installation of cameras in the market areas of the city. The third and fourth phases involved placing cameras within the entertainments area and in the area surrounding the new convention centre.

The first two phases were set up with the aid of private sponsorship and a Home Office Safer Cities grant in 1991. Nine pan, tilt and zoom cameras were installed at previously identified problem locations around the city centre core and market areas, becoming operational in March 1991. Since then, a further three pan, tilt and zoom cameras have been installed: two in the entertainment area in November 1991, and another in the town centre shopping area in the summer of 1994. Subsequent to the cameras being installed, the city centre underwent a major programme of environmental improvements including the pedestrianisation of the main city centre streets; the creation of new public squares; dismantling parts of the inner ring road; and the removal of pedestrian subways and underpasses. The police divisions responsible for the city centre were also reorganised and there was an extension of licensing hours within the city centre.

System Aims

In Birmingham, the overarching aim of the CCTV system in the city centre, and of the Citywatch Trust in general, was to 'make the city a safer place' (from Brown, 1995, p.31). According to the promotional literature, the system is used to benefit city centre users in a wide variety of ways but specific consideration is given to the:

- early detection of public disorder, anti-social behaviour and crime in order to prevent its escalation, minimise its harmful impact and aid the identification and apprehension of offenders;
- deterrence of public disorder, anti-social behaviour and crime;
- reduction of general levels of fear of crime within the town centre;
- early detection of vehicular congestion to facilitate the optimal deployment of traffic control resources;
- assistance in the general management of city life (from Brown, 1995, p.31).

In Newcastle, the main aim of the system is more specific: to support the operational policing of the city centre area. The cameras form part of a wider policing package intended to tackle burglary (including ram raiding), public disorder, theft from the person, robbery, the selling and using of drugs, traffic congestion, security and terrorism.

Operational Procedures

In Newcastle, the system is controlled entirely by the police. Images from the cameras are transmitted by microwave to four monochrome monitors located in the front desk area of the local police station where a team of police officers and civilians monitor the cameras twenty-four hours a day. The same shift system is used as for the operational police officers, thereby forming a wider operational police team. In a similar way to operational officers, the operators use the cameras to 'patrol' the city centre: searching for suspicious incidents, monitoring potentially difficult situations as they happen and keeping an eye on the local 'characters'. Although, the system records images in time-lapse mode, when required operators can also switch to real-time recording. Hard copies of images can also be produced to provide additional evidence for prosecutions. Staff monitoring the system have their own personal radios to communicate directly with the officers on the beat. The facility provides immediate communication and enables a quicker response to an incident. As the police are aware that problems are more likely at certain times and in certain places, attention is concentrated on monitoring the shopping areas during the

Figure 8.1 CCTV control room at Coventry. The efficiency of a CCTV system is crucially dependent on the skill of the operators and their ability to process a bewildering amount of visual information.

day while, during the evening and night, attention is focused on areas such as the Big Market where the majority of pubs and clubs are located.

As in Newcastle, in Birmingham the system is entirely controlled by the police. Ten monitors are housed in the main control room at Steelhouse Lane, the local police divisional headquarters. Although all the images are recorded in time-lapse mode, as at Newcastle, there is a facility permitting operators to switch to real-time recording. Civilian staff employed by the police monitor the system twenty-four hours a day working similar eight hour shifts to the patrolling beat officers. Control room staff provide the link between the camera system and the officers on the ground. A second radio link allows city centre officers, traders and camera operators to communicate with each other. As in Newcastle, the operators patrol the areas covered by the cameras, keeping an eye on local 'characters', looking out for incidents which may require a response and helping to co-ordinate the police response (**Figure 8.1**).

The Impact of CCTV

It is difficult to isolate the impact of CCTV systems. Comparison with the crime rates in adjacent areas or in similar areas where cameras have not been installed does, however, provide some indication. For comparative purposes, Brown's research study in Newcastle focused on the number of incidents recorded in four areas: the area of the four city centre police beats covered by the CCTV system; the other seven city centre police beats which surround the central CCTV area (two cameras were actually within this area); Byker (Newcastle East), one of Newcastle Central's neighbouring divisions consisting mainly of residential housing. Figures for all other divisions within the Northumbria police force area were also collated to provide an additional control measure. The periods under analysis were the twenty-six months before the cameras became fully operational in March 1993 and the fifteen months following this. In Birmingham, for the purpose of analysis, it was more difficult to separate CCTV areas and non-CCTV areas and to relate these to discrete police beats or divisions, which – in any case – were reorganised during the survey period. Furthermore, some of the area regarded as the CCTV area was not actually covered by cameras.

The Impact in Newcastle City Centre

The analysis of the results within Newcastle city centre (Brown, 1995, p.26) revealed that initially the presence of CCTV cameras had a strong deterrent effect on the incidence of a number of offences. The positive effect on some offences, however, began to fade after a period of time. The impact of the cameras varied with the type of crime.

For burglary, although there was no change in the number of burglary incidents in Byker and the rest of the force, the numbers of such incidents in *both* the CCTV area and the non-CCTV area dropped significantly. A similar crime pattern was observed in the number of criminal damage incidents. The figures also showed that for the CCTV area, the non-CCTV area and for Byker, the numbers of burglary and criminal damage incidents were all declining before the cameras were installed. After the cameras were installed the rate for these incidents fell dramatically within the CCTV area. Although, there were also reductions in the non-CCTV area, these were more gradual, especially in the case of burglary. The fall within the CCTV area occurred after the cameras were installed but, significantly, before they were fully operational. At the time of Brown's report (1995), the effect seemed to be being maintained.

The number of vehicle crime incidents dropped in all three areas, with the most marked reduction within the CCTV area where the average monthly numbers for both theft of and theft from vehicles almost halved (numbers for these incidents were, however, small). Prior to the installation of cameras, thefts of and from vehicles were declining in all areas. After the installation of the cameras, theft of vehicles continued to decline sharply within the whole of the central area whereas the level for Byker appeared to stabilise. In the CCTV area, the effect faded after eight months and after September 1993 the trend for the CCTV area became similar to that in other areas. There was a similar but weaker pattern for theft from vehicles; the only difference being that the effect occurred from the time of installation while that for thefts of vehicles came after the cameras became operational.

Compared with burglary, the effect of the cameras on juvenile disorder incidents and 'other' thefts was more difficult to discern. Incidents of juvenile disorder were increasing sharply in the central area prior to the installation of the cameras. When the cameras became fully operational, juvenile disorder incidents fell very sharply in the CCTV area. Despite a sharp increase in such incidents prior to Christmas 1993, the figure continued to fall. There was also a sharp decrease in juvenile disorder in the non-CCTV area, although curiously the timing of the reduction did not coincide with either the installation or operational use of the cameras. Despite a rise in the number of offences prior to Christmas 1993, the number of other thefts in the CCTV area also decreased but, as Brown (1995, p.22) noted, evidence for the effect of the cameras was weak.

In terms of the risk of arrest in the CCTV area, arrests and incidents of theft of and from vehicles dropped by similar amounts indicating that the risk for those offences had remained more or less stable. Despite a small drop in the number of offences following installation of the cameras, the number of arrests for drunken offences – and, therefore, the risk of arrest – increased sharply.

The Impact in Birmingham City Centre

In Birmingham, Brown (1995, p.46) noted that the research indicated that crime had reduced in those streets where there was a good CCTV view.

The presence of cameras appeared to have some effect on the incidence of robbery and theft from the person within the zone where most of the cameras were located. Prior to their installation in January 1991, changes in the rates for robbery and theft from the person for all zones were very similar. Following the installation of cameras, the incidence of robbery and theft in areas surrounding the camera zone increased sharply, and by the end of the study period, the number of offences per month was three times higher than when the cameras were installed. This pattern however was not repeated for other offences targeted by CCTV.

Rates for criminal damage in the camera zone and for the rest of the police division showed very similar rises over the study period. The rates for wounding and assault remained fairly stable in all areas after the cameras were installed. Due to the pedestrianisation schemes which changed the general context during the study period, the situation regarding vehicular crime was highly confused and no clear conclusions could be drawn.

The research at Birmingham also included a survey of members of the public. The findings of this also indicated that there was very little change in the general fear of crime amongst those interviewed or their feelings of safety within the city centre during the day. Brown (1995, p.43) did note however that there was an increase in feelings of safety for respondents using the city centre after dark amongst those who were aware that the cameras had been installed.

From the evidence gathered in Birmingham, the presence of cameras had a distinct but complex effect on the pattern of local offending. Within the city centre area, rather than reducing their overall incidence, the system acted to curb the increase in certain types of offences, namely robbery and theft from the person. Although crime in those areas with good camera coverage was lower than that elsewhere, Brown (1995, p.45–46) did note that there was considerable evidence that offending had became more common in those areas where there was little or no coverage.

As Brown (1995, p.46) notes the failure of the camera system to reduce directly overall crime levels within Birmingham city centre should not detract from the system's other benefits. The system helped police officers working within this area deal with many problems, most notably a wide range of public disorder/public safety problems. Indeed, in both cities, Brown (1995, p.62) noted how the camera systems help the police manage their resources more effectively, their primary use within city centres being as a tool to patrol areas more effectively and discover incidents as they occur. The information provided by the cameras was also used by the police to co-ordinate suitable responses to the incident and provide evidence to direct the investigation of an offence and enable the conviction of the offender, thereby saving police time and public money.

In both city centres, Brown (1995, p.62) noted that the cameras are most often used to maintain public order by dealing: 'with conspicuous anti-social and criminal behaviour, most notably various small scale public order problems, ranging from unruly nuisance behaviour to fighting and assaults'. Although the research did not show any significant reduction in this type of crime, the cameras do enable it to be dealt with more quickly. In Newcastle, for example, although the number of public disorder incidents remained unchanged after the installation of cameras, it was recognised that the strength of the system lay less in preventing these offences – which, it was argued, would occur regardless – than with co-ordinating a quick, effective response, thereby enabling officers either to defuse the situation before it became serious or to reduce the harm done.

IMPLICATIONS OF CCTV SYSTEMS

While recognising that CCTV does have some uses and can be associated in certain circumstances with reduced crime, Graham *et al.* (1996, p.16) argue that it also has several worrying aspects. They highlight four in particular: displacement effects; threats to civil liberties; threats caused by rapid advances in surveillance technology; and the threat that an emphasis on CCTV may lead to the neglect of broader and longer term policy options. To these may be added a fifth – the depersonalisation of policing.

1. *Displacement*

In terms of displacement, the evidence from Newcastle and Birmingham presents a mixed picture. In Newcastle for all the offences examined, Brown (1995, p.26) observed that there was little evidence to suggest that crime had been displaced to other locations or from one type of offence to another. There was, in fact, some evidence to suggest that there had been some 'diffusion of benefit' to the non-CCTV area, especially for criminal damage and burglary offences. In Birmingham, the research indicated that offending had increased in areas where there was partial or no camera coverage, suggesting some locational displacement; this was most evident for robbery and theft from the person (Brown, 1995, p.46). It was unclear however how far the increases in these offences in surrounding areas were a direct result of crime displacement, or of an increase in opportunities within these areas or due to a general increase within the city. It is possible that the extensive redevelopment that had taken place within areas outside the central zone – which *inter alia* resulted in an increase in the number of entertainment venues – may also have increased the number of potential targets for these offences.

Although the case study examples do not show any significant displacement, it has been observed in other locations where CCTV systems have been installed. Consideration of displacement should not, as Barr & Pease (1992, p.199) note, detract from the positive achievement of reducing crime in a particular location. To understand displacement more fully it is necessary to ask offenders if they were aware of the presence of CCTV and – if they were – whether or not it affected their decision to commit the offence (Short & Ditton, 1996, p.14). It is generally more appropriate to discuss deflection rather than displacement and to ask whether the deflection is more or less harmful than the original incidence. On the positive side, deflection might create safer corridors and enclaves but at the potential expense of more crime elsewhere.

2. *Civil Liberties*

Ostensibly employed to reduce crime and the fear of crime, the civil rights group Liberty argue that CCTV 'touches on a wide range of civil liberty issues including privacy, free association, and the democratic accountability of the police and other institutions' (Liberty Briefing, 16th October 1989). Graham *et al.* (1996, p.17) note that while CCTV systems have not yet given their controllers: 'the power of all-seeing, Orwellian "Big Brother", they may support the emergence of a large number of "Little Brothers"'. CCTV systems can give relatively unregulated individuals or agencies considerable and largely invisible powers to decide who merits closer scrutiny and control, and who has free and unhindered access to an area. Such powers may inevitably be based on their prejudices about appearances and associations rather than by evidence. Nevertheless, emphasising the apparently benign effect of technology, Davies (1996, p.xii) notes that:

> The Big Brother society imagined by the world in 1970 depended on coercion and fear. The society we are developing is more Huxley-like than Orwellian. It is *Brave New World*. Instead of the repressive tyrants and their omnipresent intrusive technology, we are witnessing a process of mass pacification.

Public concern about the infringement of civil liberties may also depend on the type of offence being detected and prosecuted through the use of CCTV. Davies (1996, p.177), for example, notes how in Kings Lynn a system originally installed to deter burglary, assault and car theft has also been used to combat meter evasion in the town's car parks. He also notes that CCTV cameras are particularly effective in detecting people using marijuana and other substances. The spate of videos depicting scenes of public violence, drunkenness and other public 'misbehaviour' culled from CCTV tapes and released prior to Christmas 1995 may also have altered people's perceptions of CCTV. Whether amusing or horrifying, such videos will have brought home to people the potential for the infringement of their own civil liberties.

3. Technology

Advances in surveillance technology also raise important questions about the possible use of CCTV systems. If the currently local CCTV systems were ever integrated into a national system and co-ordinated with a national photo ID card or driving licence system, then the capacity for central control is further increased. The increasing ability for miniaturisation of cameras – witness the camera in a cricket stump – enables increasingly covert surveillance, while many CCTV systems already have the capability to track the movements of individuals: the tracking can continue even when a person moves from one camera zone to another (Davies, 1996, p.187).

There are already increasing pressures to improve the quality of the CCTV installation (for example colour pictures and an audio facility) to provide more reliable evidence and ensure convictions. Surveillance systems have also been developed which enable computers to make facial recognitions. Known as Computerised Facial Recognition (CFR), the systems convert any face into a sequence of numbers. Using CFR, computers can 'scent through' a crowded scene to match up an individual on screen with a photographic record. To make a positive identification, evidence from more than one camera is usually required. In Britain, a CFR system has already been installed at Manchester's Maine Road football ground. The system is intended to detect known football hooligans automatically, but, as Davies (1996, p.196) warns, the application could soon be more universal.

4. Wider Policies

Towns and cities might be blinded by the rhetoric of a 'miracle technological cure' for the malaise of crime into over-optimistic expectations of the likely impact of CCTV systems. There is also the risk of a municipal 'inferiority complex' as towns and cities without CCTV systems feel vulnerable and exposed if adjacent cities have systems. Thus, a further danger is that too great an emphasis on – and faith in – CCTV systems may lead to the neglect of other, perhaps better, policy options. In this respect, CCTV should not be considered in isolation but ought to be integrated with other measures, such as mixed uses, policing strategies, Shop Watch, Car Park Watch, Pub Watch, design improvements, etc.

5. Depersonalisation of Policing

CCTV systems seem to have a generally beneficial and positive effect on crime detection and the prosecution of that crime. They also appear to have a positive effect on the incidence of crime within the area surveyed. The effect on crime prevention and people's fear of crime is, however, unclear. Perhaps the most positive benefit of CCTV systems is to enable the police to deploy and utilise their resources more efficiently and effectively. Regardless of any other effect, this is a significant contribution to a safer city centre. What is less advantageous however is the anonymity and depersonalisation that results as the police are further distanced from the population they serve. In the UK, some towns no longer have a police presence at pub closing-time – the police preferring to react only to what they see on their monitors (Davies, 1996, p.199). This is at a time when there seems to be a desire for greater police visibility. Indeed many surveys indicate that people express a desire for police presence (for example Box *et al.*, 1988; Guessoum-Benderbouz, 1994).

Furthermore, the presence of surveillance cameras may itself be a fear generator; rather than reassuring citizens, it may alarm them into thinking that – if cameras are needed – there must be a serious crime problem or, for those within the defended area, it might create a 'siege mentality' or that areas not surveyed by cameras must – by definition – be unsafe.

CONCLUSION

CCTV technology is here to stay. While the technology ought to be tamed and used to the benefit of society as a whole, it is by no means certain that it will. The critical issue is how society chooses to use it; how it monitors that use; who uses it; who watches the screens; who has access to the tapes and when the tapes are erased; and what safeguards there are for the protection of individual freedoms and civil liberties. Expressing concern regarding civil liberties and the intrusion on privacy, the Local Government Information Unit (1994) report, *Candid Cameras*, called for a code of practice for CCTV schemes, noted that CCTV was not regulated by legislation and that additional care was required on the part of locally accountable bodies. The LGIU (1994) also noted, for example, the potential for CCTV to have a 'chilling effect' on otherwise legitimate activity, such as trade union demonstrations outside the town hall. Thus, what is currently a matter of simple trust needs to be replaced by rules and safeguards (Davies, 1996, p.201). Furthermore, as Beck & Willis (1994, p.175) argue: 'The transparency of the controls, as much as the controls themselves, is crucial to the reassurance which is required'. Only with sufficient and transparent regulation and safeguards on their use will concerns about the erosion of civil liberties and of the democratic public realm be alleviated; thereby enabling CCTV systems to metamorphose from a threatening 'Big Brother' into a more benign 'Big Father' (Stansfield, 1995). In addition, it should supplement rather than replace other forms of surveillance, including formal surveillance by the police.

9

Safer Transport and Parking

Safe cities need safe transport. If city centres are to be made safer and more attractive, then getting in and out of them and moving around has to be and to feel safe. At the moment this is not so. Survey after survey shows that the problems of travel, especially at night, are a serious deterrent to a whole range of recreational and social activities (**see Table 2.4**). This is particularly – but not uniquely – a problem for women. It applies not just to those who travel on public transport, but also to private car and even taxi passengers.

All forms of transport have been following policies which contribute to this problem for at least twenty years. More and more railway stations are now unstaffed halts; traffic is given priority over pedestrians who are made to use subways instead of crossing busy roads at street level; conductors have largely disappeared from buses; multi-storey car parks use pay-and-display machines instead of cashiers, and a private hire taxi cab industry has grown in many areas without a system of strict vetting and licensing. As public transport declines and city centres are less well-used at night, so bus stops, rail and bus stations and car parks all become lonelier and therefore more uncomfortable places.

This is both the best of times and the worst of times to be looking at the problems presented by making transport safer. It is a very opportune time because current concerns over the financial costs of road programmes, and the environmental costs of increasing car use in urban areas, focus attention on public transport and how to make services good enough to attract people out of their cars. Similarly, the light rail systems (LRT) introduced in Manchester and Sheffield and under development in Nottingham are being constructed at a time when planners are more aware of safety considerations. The consultants for the Nottingham LRT for example, have shown their plans at workshops with women's groups to find out how far they meet women's concerns over safer travel. Moreover, policies to preserve and revitalise city centres are changing attitudes to multi-storey car parks and bus route penetration of city centres.

However, the financial and institutional conditions could hardly be less favourable to the kind of changes that are needed. It is a bad time to talk about any increase in public expenditure; local authority budgets are being cut not

Copyright © 1997, Sylvia Trench

expanded, and any revenue released from cutting the road programme is likely to help finance tax cuts rather than improvements to public transport infrastructure. Organisationally, public transport is in a mess, with deregulation making it difficult for local authorities to influence the behaviour of bus companies. The fragmentation of the railway system into ninety different companies and a separate one for the track will make it difficult for them to deal with major new improvements to station and train safety in the near future.

This chapter reviews the possibilities for reversing some of these trends to make travel to and within cities a less threatening experience. It draws on the views expressed by representatives of women's organisations taking part in a series of workshops at the University of Nottingham.

FEAR AND TRAVEL

Surveys in a number of UK cities establish that around two thirds of women are afraid to go out at night alone. Significant numbers will not use public transport and are worried about city centre car parks (Atkins, 1989). In Bradford, a survey found that as many as 59 per cent of women avoided using any form of public transport at night (Local Transport Today, 1990). In Nottingham as many as 54 per cent of women interviewed on two low income housing estates said they never used buses after dark, and a further 30 per cent only used them rarely. These were areas where only 21 per cent of women had a driving licence. Seventy-eight per cent reported feeling 'not at all safe' or 'not very safe' waiting at bus stops, and almost as many felt as insecure walking to the bus stop (Hiley, 1995).

Women take a variety of measures to avoid situations seen as dangerous: they avoid walking alone at night, or only go out if a safe return has been arranged in advance; they avoid unsafe areas like subways and back streets and waiting at bus stops (see Chapter 2). Fear of crime increases the use of cars and the demand for close-by parking (Citizens Crime Commission, 1985).

PUBLIC TRANSPORT

City centres could be made more accessible by putting money and effort into measures which improve public transport so that people can feel safe and comfortable travelling on ordinary services. The kind of measures involved – increased staffing at stations, conductors on evening buses, hail and ride minibus services penetrating housing estates – provide other kinds of benefits as well as improving safety, and assist other groups beside those worried about attacks. However, they require increasing public expenditure in a sector that is already underfunded and under pressure, and, even if the money were to be found, the public system would still not be as secure as a door-to-door service.

This is why a number of local authorities and voluntary groups have sought to provide special safe women's transport services to offer a segregated door-to-door service for women who pre-book special minibus facilities. Many have been set up by local authorities often working with voluntary groups[1] and more recently as part of the Home Office's Safer Cities initiative.

Women's Transport Schemes

The first segregated women's transport scheme, established in Bristol in 1988 after a 23 per cent increase in reported rapes and other violent attacks, is a good example of this type of service. It provides a door-to-door evening lift service for women who cannot afford or who do not feel able to use other forms of transport at night. It uses volunteer drivers and runs two vehicles, both adapted for wheelchairs. In its first year of operation its 200 members made over 4,000 trips. Unfortunately, the limited number of vehicles and drivers has meant that they have had to restrict use of the service, giving priority to women on low incomes, black and ethnic minority women, disabled, elderly and young women and those with a particular fear of violence.

The Homerunner service in Bradford, started in October 1989, targets similar groups to those in Bristol, with the addition of female shift workers and those wishing to attend evening classes. It was set up as a door-to-door service for women only, operating three minibuses between 6 pm and 11 pm Monday to Saturday. Users pay a 90 pence flat fare which covers about one-third of the costs, with the deficit being covered by a £45,000 per annum subsidy from Bradford Safer City Project and the West Yorkshire Passenger Transport Authority. The service has carried between 300 and 400 passengers a week, coming predominantly from households without a car. Research shows that 60 per cent of journeys are to and from work and around 60 per cent of the leisure trips would not have been made without the service. Employers have shown interest in the service for night-time travel by their female employees, and offers of sponsorship are being investigated (Local Transport Today, 1990).

Other safe transport schemes have been running at one time or another in a number of British towns and cities including: Brighton, Stockwell, Hammersmith & Fulham, Manchester, Preston, Leeds, the Wirral and Nottingham.

Safe transport schemes clearly meet a genuine need. At the moment, a few projects operate over relatively small areas. They can only cater for a fraction of potential demand and require pre-booking. There is little doubt that those who use them are very conscious of significant benefits despite the fact that they usually need to be booked well in advance. But it is unlikely that they are able to spread their benefits right across the community and may become concentrated on a small group of regulars. For example, Merseyside PTE's Homesafe service on the Wirral peninsula was one of the most apparently successful of these schemes, but was discontinued after eighteen months because it was felt that the £60,000 subsidy would produce more benefits if it were used to promote safer travel generally rather than a high quality service to its relatively small group of regular users.

1 Special services for women could run into difficulty with equal opportunities legislation but this can be avoided if funding is channelled through voluntary bodies.

The big question for those concerned with women's mobility is whether to go for the expansion of such schemes as the main way of dealing with the general problem of safety across the whole spectrum of public transport services for all women. This would appear to have two big dangers. First, by perpetuating the notion of a curfew for women it may actually contribute to increasing their fear of crime, and discourage them even more from using public transport. This in turn leads to the second danger: if such services were run on a scale to make a significant impact on transport for all women, then the numbers of people using public transport at night would fall to even lower levels and increase the financial pressures for economies in staffing.

At present, the numbers using safe women's transport schemes are nowhere near the kind of levels which could threaten public transport and cause these longer term draw backs. In the short term, the services provide a lifeline to the very vulnerable women who use them and, in cases where women have been intimidated into not going out at all, are often the first step to changing their attitudes to going out and about. There is an argument for all local authorities to consider supporting such schemes on the same basis as they support community buses for low revenue routes in rural areas, so that some safe transport scheme is available in all areas.

Whatever happens about safe women's services, the main sources of anxiety in the general transport system have to be removed. Thanks to the work of women's committees in the old Greater London Council and the Metropolitan Counties, Home Office Safer City projects and the work of organisations such as Crime Concern who have undertaken long term studies with a number of major transport undertakings, such as Merseyside PTE, Centro, etc., there is now a good understanding of what are the perceived problem areas and what kind of remedies are practical and useful. The following sections review what should and is being done in relation to bus and rail services, subways, car parks and taxis.

Bus Services

Bus stops are a major cause of complaint. They are often located in lonely places and moving them nearer to shops or petrol stations where there is some evening activity is a simple measure which costs little (Atkins, 1989). Bus stations are generally insalubrious places – like many railway stations, the worst part about them is the lack of a staff presence. One of the good results of deregulation is that more buses circulate around the central parts of town instead of being relegated to the bus station and this makes it easier for people to choose a waiting place.

There is a lot of scope for improving bus shelters by better lighting, provision of timetable information and telephones. There is some disagreement about whether providing seats is a good idea – some women consider that seats actively encourage gatherings of young people which can be intimidating.

Information about how long people need to wait for a bus is clearly helpful, and some cities, such as Nottingham, are introducing so-called Real Time systems. Nottingham now has thirteen bus stops equipped with visual displays showing the arrival time and destination of up to the next six buses.

Double decker buses worry female passengers at night. Women say they feel out of sight and far away from the driver. Operators do try and avoid using them at night but some, such as Nottingham City Transport, have so many double deckers in service that there is often no alternative. One less expensive expedient would be to follow the example of Barnsley where the upper deck is cordoned off at night on less busy services.

Education and training are both important – drivers in particular need to be educated about behaviour and about body language which they might think friendly but nervous women might find intimidating. Drivers also need to be trained to treat children sensitively and not refuse to carry them when this would put them at risk (Beuret, 1994). Many bus companies, such as Trent and Nottingham City Transport, are now training drivers, and examining why a few seem repeatedly to get into trouble. Research shows that women drivers are better at avoiding incidents, and when they are employed the number of incidents falls more than proportionately (Beuret, 1994)

Hail and Ride

Services which allow passengers to get on and off buses wherever they choose along a fixed route have a clear advantage in terms of safety, especially if they are minibus type services which penetrate housing estates and can drop passengers off close to their home. The entire Toronto Metropolitan area now has buses which stop anywhere along the route on demand after 8 pm following the success of a pilot scheme in one area. The city of New York also now has a pilot request stop programme (Wekerle & Whitzman, 1995).

Changing evening bus services to 'hail and ride' on less densely used routes has been successfully tried in Stockport. It has not required any special financing apart from some expenditure on publicity – in fact, the new service generated extra travel (see, for example, the National Association of Local Government Women's Committees, 1991). On the other hand, the Close to Home Service run by Merseyside PTE – a six month Hail and Ride experiment – was discontinued because it was not used well enough to justify its £30,000 annual cost.

Bus operators are cautious about introducing more such services – heavily loaded ones would be so delayed by frequent stops as to deter passengers, but at night with fewer passengers they would seem to have a lot to commend them. Many operators seem happier with the idea of setting down on request than with full hail and ride which would also pick passengers up anywhere *en route*.

Some bus operators admit that some services especially on inter-urban routes are effectively Hail and Ride in the evenings. Drivers drop passengers off on request on an unofficial basis. This is apparently very common practice for regular passengers on late night services (IPS, 1994). However, this is presumably only known to regular late night passengers who have had the courage to test the system. It would be nice if bus companies could move on from this unofficial practice to publicising a guaranteed service which set down on request after a certain time of night. Passengers might be willing to pay an extra charge for Hail and Ride services which picked up and set down at points on the route, or even those which just set down near home if the 'hail' part was not practical.

Conductors

Most people would welcome the reappearance of conductors on buses but conductors do not significantly increase overall usage. Bus operators in the Midlands who have experimented with introducing them on selected services argue that it is hard to justify the 80 per cent increase in operating costs for a relatively small rise in usage, especially if no money is available for improvements to other parts of the journey (Institute of Planning Studies, 1994). Public transport is judged by the quality of the journey as a whole and, if the walk to the stop and the wait are felt to be unsafe, then having conductors on the bus does not deal with the rest of the problem.

One example of a pioneer attempt to deal with the whole journey is the Safer Bus Routes project in the St Ann's and Sneinton areas of Nottingham. This is a joint venture between the County Council, the City Council, City Challenge and a local bus company for a concentrated set of improvements on and off route. It involves improving both the service and access to it. Bus frequencies would be increased; regular drivers would operate at the same times; bus shelters and the routes between nearby housing and the bus stops would be well-lit; and shelters would have phones and timetable information. It is intended as a pilot scheme which may provide a model for other areas of the city. A feasibility study (Hiley, 1995) to find out how far potential passengers might be attracted back by such facilities confirms that:

> Fear of crime is clearly an impediment to bus travel for a high proportion of women and vulnerable sections of the community in Sneinton and St. Ann's. The point at which fear is at its highest during journeys is during the wait at the bus stop, especially at night. The journey to the stop on foot is also a major concern.

Between 40 and 54 per cent of women said they would use buses more often if they felt safer. The main changes they sought were: better lighting on the way to and at bus stops; bus shelters with phones and large clear timetables; and buses with two way radios which could be used in emergencies to help them or relay timetable problems via the phones in bus shelters. EU financial support is being sought for the projects but bus operators are optimistic about their commercial prospects.

These pilot schemes illustrate the problems of getting finance for expenditure which has important social benefits but may not be commercially justified. In this case, finance for the expensive added-safety features, such as lighting and bus shelters, will come from outside sources – in this case, City Challenge and perhaps EU. As this kind of subsidy could provide such significant benefits, it seems a pity that one of the 'competition' provisions of the 1985 Transport Act limits the amount of subsidy that can be paid to any one operator for this kind of project (Transport Act, 1985)

Local Rail Services

Unstaffed stations at night are a major cause of worry. Whenever travel problems are under discussion, staff presence especially at night is mentioned as the single most important contribution to making travellers feel more secure.

A number of cities in Britain are in the process of planning new light rapid transit (LRT) systems. It is at this stage that plans should be made to provide secure travel with well-sited stops. The local authorities and their consultants responsible for the new LRT system for Nottingham used the workshop at the University of Nottingham Conference in 1994 to consult women's groups on its design (McClintock, 1994)

The main message from potential women users is that stations should be staffed at all times – it is not enough for operators to promise some roving personnel. Some say they would be prepared to pay higher fares if staffing could be guaranteed, especially on late services where there is a greater chance of rowdy and unruly behaviour. At such times, there are fewer other passengers around and people feel uneasy. Extra staff could also help passengers with onward travel arrangements at the end of their LRT journey. Having staff would assist all travellers, including children, the elderly and disabled, not just women. It is generally felt that it is not enough to rely on drivers and alarm systems in cases of emergency.

Many systems now use Closed Circuit Television (CCTV) for monitoring stations. Clearly this is better than nothing, but most women say that it does not make them feel any safer and some even mention a fear of voyeuristic youths at the monitors. Some rail systems have found less expensive ways of compensating for feelings of isolation.

Opinions differ about the merits of having separate rail compartments for women passengers. Muslim women, who may be reluctant to sit next to strange male passengers for cultural reasons, appear to welcome them, but others argue that separate compartments would encourage the idea that public transport was not safe for women to use. They prefer the open interior design, with long views and good lighting. This benefits all users and, moreover, allows men to help women with children and pushchairs (McClintock, 1994).

Passengers worry about getting change from automatic ticket machines and possibly being thrown off the system at isolated spots and awkward times. Nottingham LRT hopes to avoid these problems by arranging for tickets to be purchased through a wide variety of outlets, and eventually by the use of pre-charged smart cards.

Examples of Good Practice

On some stations, London Transport offers passengers the opportunity to wait in the well-lit station entrance with the ticket collector until the train is announced. The New York city subway system also has a policy of designated off-hours waiting areas (DWAs) on the platform in view of ticket collectors. Toronto Transit Commission has these DWAs on all its platforms, boldly-signed, brightly-lit, surveyed by TV cameras and with emergency buttons and intercom linked to security staff. These are located where the carriage staffed by an attendant will stop (Wekerle & Whitzman, 1995).

Tyne and Wear Metro runs shorter trains in the evening to concentrate their passengers and avoid empty compartments. In West Yorkshire, the Metro has a working party with its own budget for safety and security which advises on design of new stations. They have a policy of replacing subways with foot-bridges and of providing phones in stations. There is also liaison with local planning departments to get compatible adjacent uses to otherwise deserted stations which will bring evening life and activity to the vicinity. They use CCTV linked to a central control room.

The Crime Concern Consultancy developed multi-agency partnerships between the London Underground, local authorities and businesses for three pairs of stations as pilot projects to see if they could work together to improve the areas in the vicinity of stations. They also worked with Mersey Travel on similar partnerships for Kirby and Birkenhead. Merseyside PTE have a strong financial and policy commitment to safer travel. Following two years' work with Crime Concern on improving safety, they now have all bus and rail stations staffed at all times of day and are building six new bus stations which will also be staffed. They employ forty British Transport police to rove over their network. They are also leading the Travel Safe Programme of the Merseyside Police Authority Safer Communities Partnership. This has attracted funding of £500,000 over five years from the Single Regeneration Budget for a variety of safety-related improvements to public transport, such as provision of two-way radios on buses, CCTV, and staff training and education on safety matters.

Strathclyde PTE have police roving on its network. South Yorkshire PTE use posters asking people to report incidents and CCTV to chase them, staff all interchanges and employ uniformed security staff at bus stations at night. They also put phones in bus shelters, clean graffiti promptly, and use two way radios and videos on buses.

Car Park Design

Until recently, planners were very committed to segregation to avoid conflicts between pedestrians and vehicles and they designed highways so as to ensure minimum interruption to vehicular mobility; inner ring roads were crossed by pedestrian subways and cars were left at multi-storey car parks built on the fringes of the city centre.

Policies towards car parking need to be reviewed in the light of safety considerations. Car parks are often designed for the car not the pedestrian – this makes them inconvenient for all drivers – and a lack of safety features makes them greatly feared by women. Car parks are often located on poorly-lit side and back streets where queuing cars will not impede traffic. They are often separated from the centre of pedestrian activity by roads accessible only via subways or across busy roads. Although car parks linked to shopping centres provide quick and easy access by day, at night the accessways may be closed leaving pedestrians to cross surrounding roads (Jones, 1991). For these reasons, it can be safer to enter the city centre by bus if the bus stops are on busy streets.

Parking structures are often considered to be unsightly and visually intrusive, and planting or physical barriers are used to 'camouflage' them. This might improve the visual amenity of the area, but it eradicates the possibility of informal surveillance from surrounding housing, streets, shops and offices (Jones, 1991). A lack of natural surveillance opportunities in parking areas not overlooked by housing or busy shops and offices increases the scope for vehicle crime. Many argue that car parks should be overlooked by surrounding buildings and pedestrian activity (for example Webb *et al.*, 1992).

Multi-storey car parks present many problems – they are normally unstaffed, visibility is poor, and lifts and stairs for people arriving and leaving are out of sight of traffic both in and outside the car park. Design improvements which increase visibility are possible as shown by examples in Nuneaton and Leicester (Atkins, 1989). The car park design guide produced by Birmingham City Council (1992) incorporates many safety features. The guide is based on a questionnaire survey of over 500 women on car park safety and design, undertaken by the council's Department of Planning & Architecture (Birmingham City Council, 1990).[2]

The development and adoption of improved designs has been encouraged by the work of ACPO and the Automobile Association (AA) who together run a joint scheme giving prizes for safer designs. The successful schemes seem to involve providing attendants and introducing CCTV. For example, Nottingham's Broad Marsh car park was recently revamped and won an award. The improvements have produced a striking reduction in crime figures – but women are still not very happy and argue that the main benefits have been in terms of property-based crimes and vandalism. The stairs are still concealed and lifts are scary.

The presence of staff has been shown to reduce crime and increase feelings of security and some women favour the designation of women-only car parks or the provision of ground floor spaces, near the attendant, reserved for women. This system works well in a number of German cities but when Birmingham City Council tried to introduce a small central area women-only car park it was advised that this could contravene equal opportunities' legislation and the proposal was shelved. Women's groups consulted about women-only car parks without attendants were divided about whether they should be supported (Trench *et al.*, 1992).

[2] There are also a number of research reports on car park design. See for example Poyne (1992); Webb, *et al.*, (1992); Webb & Laycock (1992); Laycock & Austin (1992) and Tilley (1993a).

A less ambitious experimental scheme in Exeter in 1996, where ten spaces in a well-lit, convenient area of a multi-storey car park are marked for women users, may survive despite the council being given similar advice from the Equal Opportunities Commission. It would be interesting to see whether the EOC would be able to justify spending between £1,000 and £2,000 per day to support a male objector in the County Court in the event of a council such as Exeter sticking to its very modest pro-women gesture in the face of objections. Exeter is probably protected from this kind of legal test as long as it does nothing to enforce the 'women-only' signs. If the signs are observed, despite their status as a request to male drivers, the scheme may well survive and be expanded after its trial period, despite some local controversy.

A more substantial women-only provision is being successfully operated in Cardiff by the ACPOA company with the backing of Cardiff City Council. Forty-two spaces near the main office on level one of the six floor Knox Road car park were sectioned off for women. This provision has now been extended to the whole floor, and together with other safety features, such as CCTV, good lighting and frequent security patrols, helped it win the Norwich Union's Safe Car Park award in 1994. As in Exeter, the company is relying on the voluntary status of its designated spaces to avoid action under equal opportunities legislation. The company has plans to extend the scheme to other British towns and cities including London. The Forest of Dean District Council are also using segregated provision for women. In this case, the council responded to a tragic attack on a female member of staff in Gloucester by designating a staff car park at the rear of the main office as exclusively for women until 4 pm. They invite other members of staff who feel vulnerable to apply for permission to park there, and assume that this provision will also spare them action under equal opportunities legislation. It is seems, therefore, that British equal opportunities legislation need not prevent local authorities from offering women-only facilities with extra safety features. Now that a few have set a precedent, perhaps more will follow their example.

Most people consider well-lit, attended open air parking lots to be safer than multi-storey car parks. In recent years, traffic calming measures have made it possible for people and vehicles to share the use of central areas with increased safety for pedestrians; the next developments need to be allowing increased on-street parking at night and improving staffing in central area car parks. There is a case for allowing on-street parking in the city centre, especially in the evening, and local authorities should consider this. Workshops with women at the University of Nottingham found that most people considered on-street parking to be the safest form of parking, especially at night. Many people who have access to cars say that they would be more willing to come to city centres if they could park on the street close to the wine bars, restaurants, theatres, etc., that they use. Although on-street parking is safer in terms of personal safety, the car itself may be more vulnerable to theft. Research for the Home Office, for example, suggests that the most vulnerable vehicles are those parked in public places with little security (Houghton, 1992). Although it is desirable that both people and property are safe, on-street parking would at least allow individuals to decide their own trade off between personal and property safety.

Subways

Pedestrian subways and barriers to stop people crossing at street level were part of a policy dating from the 1960s which valued motorists' time more than the convenience of elderly and laden pedestrians. Nowadays pedestrians are given more consideration. Safety fears have added another reason for revising the policy. Women at the University of Nottingham's workshops have been unanimous in rejecting CCTV as a way of making subways acceptable. There are ways of improving their design (see Atkins, 1989) to increase visibility or use shops to make them less deserted, but complete abolition is generally preferred and brings benefits to all those using the city on foot. Nottingham City Council is replacing subways with pedestrian surface crossings, as will a number of other British cities including Sheffield, Portsmouth and Poole.

Taxi Services

Taxis might be the ideal way for women without a car to have a door-to-door service, and women's use of taxis is increasing even among lower income groups (Trench & Lister, 1990). However, a number of recent cases of attacks on women passengers inhibit this growth and there are many fears about them. If a woman calls a taxi which fails to turn up, she could be stranded in an unsafe place or be dependent on hailing an unknown cab. The use of the London style 'Black' taxi is a safeguard for women because they can recognise a licensed vehicle and because it is safer to travel in. Nottingham is one of a number of local authorities that are making its use compulsory for licensed taxis.

More needs to be done to increase the reliability of taxis and the vetting of their drivers. It is unfortunate that the development of women-only cab firms driven by and for women is, like the Birmingham car park, running into trouble with equal opportunities legislation.

Vetting procedures also need to be tightened up even though there have been some recent improvements. It is still possible for a convicted rapist to get a licence to drive a cab after enough time has elapsed (IPS, 1994). Passengers often fail to make a formal complaint after unfortunate incidents and more could be done to make people aware of the importance of complaining to the licensing authority and how to do it (see, for example, Peter Perkins' comments, IPS, 1994). The Taxi Licensing Officers Association is encouraging firms to produce a leaflet for the public and is considering distributing it with rating notices.

Local authorities might do more on this issue, for example, by promoting wider application of something like the voluntary taxi operators code introduced by the London Borough of Hackney (NALGWC, 1991). Drivers are carefully vetted and trained to be courteous and to avoid suggestive behaviour.

Woman callers are given the name of the driver who will meet them; the driver will also know their name. The White Paper on Taxis (Department of Transport, 1995) has recommended training for taxi drivers. This would clearly be beneficial but excludes the private hire sector. Most lay people are not aware of the distinction between taxis licensed to ply for hire on the streets and at ranks, and private hire cars which should only respond to pre-booking. The degree of vetting and licence requirements for the private hire sector varies between local authorities and, in many areas, needs to be tightened up.

Safety Audits

Toronto set up a special action committee on safety of the public transport system following its adoption of *Municipal Strategies for Preventing Public Violence Against Women* in 1988 (City of Toronto, 1988) which pioneered the use of safety audits and the participation of women's groups in identifying unsafe places (**Table 4.2**). The idea has been taken up by groups in other countries – for example, in Nottingham STRIDE, a pressure group for safer travel for women, has produced a manual for local organisations and a guide to safety audits on transport in Nottingham (STRIDE, 1995). It is simple in conception and requires a small outlay of time and effort from voluntary groups but provides an effective way of focusing public and professional attention on problem areas.

Crime Concern, an independent consultancy associated with the Home Office, has done a number of in-depth studies with some of the larger transport operators who now have a lot of information about exactly what worries their passengers. They have prepared a comprehensive report on safety policies throughout UK public transport (Crime Concern, 1995a; 1995b; 1995c).

CONCLUSION: PRIORITISING SAFETY

A number of informal women's safety networks and pressure groups have clearly had an impact on changing attitudes of transport operators. Relatively small sums of money, made available as a result initially of Greater London Council and Metropolitan Counties Women's Committees and later the Home Office Safer Cities Projects and City Challenge, have significantly raised the profile of safety issues in transport. Although they can do little to deal with the basic lack of finance for major changes in methods of running services, it is thanks to their efforts that safety issues are very much on the agenda and that a largely male-dominated transport profession has become aware of some very fundamental and long neglected needs of its female customers.

It is interesting to note, however, that the profession still shows a very stereotypical male bias towards technological rather than human remedies. For example, funding for CCTV – which does produce reductions in actual levels of crime but is not much liked by women – and for the £200,000 *real time* bus information system in Nottingham often comes partly or entirely from central government. It is much harder to get funds for putting more human beings on the streets, in bus and rail stations and in car parks. The cynical might wish that the resources of commercial companies were available to promote and advertise the benefits of a human presence as energetically as they are for the promotion of technical equipment.

The decision of the Department of Transport to publish guidelines for transport operators on *Personal Security on Public Transport* (Department of Transport, 1996) ought also to be a sign that the issue is being given some priority. Unfortunately this expensively produced folder is of little help. It is mostly platitudes and has little if anything specific to say. The transport operator who needs to be told things like 'identify the issues' or 'evaluation of success or failure is vital' is unlikely to be able to run a bus company in the first place let alone implement a safety policy. Anyone seeking to know what policies it would be useful to carry out would be better advised to read the publications of Stephen Atkins (1989) and of Crime Concern. Nonetheless, the examples of good practice in this chapter are an indication of changing attitudes. In the early years, conferences on safety issues were largely attended by women and very few transport operators showed any interest. The University of Nottingham Conference, Planning for Safety in Towns, in 1994 was actually sponsored by the Light Rail development agency of the County Council and the two local bus companies, Nottingham City Transport and Trent Buses; another indicator of a significant change in attitude and commitment.

Safety issues in transport now have a much higher profile than a few years ago and this is very encouraging. Two major problems remain: first, how to finance the kind of improvements that it is agreed are needed, and, second, how to implement the co-ordinated approach required in a system which has been so fragmented by bus deregulation and rail privatisation. There is a lot of support for safety policies at local level. Now progress depends on central government making their implementation easier.

10

Housing and Safer City Centres

In city centres, especially outside shop and office hours, residential uses can help to create a 'living heart'. The twenty-four hour life brought by residents is a crucial contribution to its vitality. More residents result in greater indigenous demand for facilities in the city centre; thereby increasing the number and mix of uses. There is a strong perception that 'peopled places' seem safer. A survey by Valentine (1990, p.288–303) showed that women feel safer in the presence of others, while Fisher & Worpole (1988, p.19–21) observe that busy streets acting as foci for social life are generally safer streets. Thus, the presence of people and their 'eyes on the streets' (Jacobs, 1961), by implication, increases the city centre's safety. For city streets to be better populated, city centre residential developments are essential.

The flight of population from the central areas of cities to peripheral suburbs during the twentieth century has left the central area of many cities with a declining resident population. Due to developments in transport, there has been a general move of population away from living in town and city centres. This was further exacerbated by Modernist planning theories and practices that sought to functionally zone cities. The 1980s and early 1990s, however, have seen evidence of a desire for urban living by some social groups. Such groups no longer want to be 'out in the sticks' but back 'where the action is'. Furthermore, it is projected that in the UK there will be a need for 4.4 million new homes by 2016. Although the population is not expected to grow significantly, this housing need results from increasing family fragmentation and declining household size. Some of these will undoubtedly go into the green belt as there is still a strong demand for suburban living. The government would, however, like to channel a significant proportion of the new homes into city centres and onto the edge of city centres. Such locations will only be desirable if the city centre is able to provide attractive and high-quality environments and if crucially the housing and the area are considered to be safe.

CITY CENTRE LIVING

There is evidence of a growing demand for central area housing – and the nascent emergence of an urban culture. Provided it has vitality, activity and –

crucially – is safe, many people are prepared to live in city centres. Furthermore as more people delay having children and with a greater proportion of active pensioners more people inhabit life stages where the vitality and life offered by city centre living is attractive. In addition, the General Household Survey shows that the most common household today is a couple with no children (36%) (from URBED *et al.*, 1994, p.13). What has also been observed is that women in general, and gay men and women in particular, often have a greater preference for city centre living. In part, this may be because women are relatively more affluent than they had previously been. Equally, it also reflects the necessity of juggling the demands of an often more complex lifestyle. There are other types of residential accommodation that would suit city centre locations, such as student halls of residence and 'foyer' schemes – an idea originating in France for 'student-type accommodation for non-students'. Furthermore, concerns about job security and a hiatus in the moral exhortation to house purchase suggest a possible revival of the rented sector – people feel that they may as well rent as buy – which may benefit city centre locations more than suburban ones.

Ashworth & Tunbridge (1990, p.115) argue that there are two principal attractions to living in the city. The first is the convenience of the location; the second is the historic character and allure of the residential space available. In Britain and the United States, in particular, the dominant image of the residential environment has been that of suburbia. Beginning in New York's SoHo around 1970, however, for certain sections of the residential market, as the bare, polished wood floors, exposed red brick walls, and cast-iron facades of 'artists quarters' gained increasing notice 'the economic and aesthetic virtues of "loft living" were transformed into bourgeois chic' (Zukin, 1989, p.2). A romance and charm – together with a certain social cachet – attaches itself to the inhabitation of former industrial spaces. As historic properties possess a scarcity value, distinctive housing has been one way of revitalising historic areas, for example, New York's SoHo and London's Shad Thames (see Tiesdell *et al.*, 1996).

Historicity and centrality may be mutually reinforcing but, as Ashworth & Tunbridge (1990, p.115) note, there is an important distinction between those who live in historic premises and those who – although resident in the same properties or areas – have a different set of motives and priorities. The first group value the historic character as such; live there because of it; invest in the properties and support its conservation. By contrast, the second group is relatively indifferent to the historical character, placing a higher priority on the area's centrality and accessibility – and perhaps most importantly – its low rents and/or flexibility of tenure. In established residential locations, through a process of gentrification, the first group will often tend to displace the latter. While not neglecting the issue of gentrification, the focus of this chapter is, however, on increasing the residential population not displacing it. (For a full discussion of gentrification see Smith, 1996.)

A further argument for more urban living is as a response to concerns about environmental sustainability. Although the need for sustainable development has been accepted, there is considerable debate over its practical implications. For planning, the most important environmental problem is the energy-intensive nature of land-use patterns. Attention has therefore focused on urban form and the contribution that different forms could make to lowering energy consumption and pollution levels. The most widely publicised concept is the 'compact city' which advocates denser, more compact urban forms as environmentally more sustainable. The compact city is essentially a rather vague concept which encompasses ideas of higher densities, more public transport and a mixture of uses. The intention is that these factors will reduce the amount and length of car trips. While in *extremis* the compact city idea is open to debate (see for example Breheny, 1993; Goodchild, 1994; Jenks *et al.*, 1996), more compact urban forms and increased residential development in the centre of towns will generally reduce the need for transport. Development in the central area might also alleviate development on green field sites and in peripheral locations.

Thus, in addition to its contribution to a safer city, there are other justifications for increasing the residential population of the city centre. Nevertheless, to have more city centre living in the long term requires an urban vitality to be developed in the short term. To ensure an urban vitality, the overarching prerequisite for city centre living is for it to be perceived as safe.

SAFER RESIDENTIAL AREAS

At the area level there is considerable debate about the merits of 'integration' versus 'segregation'. Some commentators, such as Hillier (1988), advocate the merits of integrating residential developments into their immediate surroundings. Many others, including Newman (1973) and Coleman (1985), argue the merits of enclosure. Integration is often proposed as a positive means of 'de-ghetto-ising' concentrations of public housing and of integrating them physically into the surrounding area. Segregation, by contrast, is seen as a way of insulating richer neighbourhoods from poorer neighbourhoods. Security through segregation can also be seen as a 'positional' good and as a symbol of prestige that helps define and establish social status '"security" has less to do with personal safety than with the degree of personal insulation . . . from "unsavoury" groups and individuals, even crowds in general' (Davis, 1990, p.224).

On crime prevention grounds Poyner (1983, p.36) argued that in the interests of burglary prevention: 'areas of wealthy or middle class/ middle income housing should be separated as far as possible from poorer housing'. In terms of social justice, Bottoms (1990, p.19) hopes that many people would join him in rejecting Poyner's suggestion: 'we should try to foster a sense of community and common interest among people of different social classes and income levels, rather than to emphasize social separateness by . . . residential segregation'.

Segregated Residential Areas

Despite Bottoms' sentiments, the popular tendency is towards segregation. Sorkin (1992, p.xiv) notes how, in the United States, the impulse towards segregation is ubiquitous: 'city planning has ceased its historic role as an integrator of communities in favour of managing selective development and enforcing distinctions'. Segregation with the possibility of exclusion prevents non-residents from entering the area and also reduces their 'awareness' of the area. Through their daily routines, all people – including those who commit crime – develop an 'awareness space' from which criminal targets are usually selected (Brantingham & Brantingham, 1991; 1993; see also Chapter 2). If 'those who commit crimes' can be identified and excluded, then the residential area will not constitute part of their awareness space and will arguably be safer. If it is not possible to identify those who commit crimes precisely, the same logic applies if most people are excluded. The approach is risk averse and it is less risky to exclude too many rather than too few.

There are various forms and degrees of segregated residential areas. They often involve the situational crime prevention techniques of target hardening, access control and entry/exit screening practised at the meso-scale, thereby creating a 'fortress' neighbourhood. Oscar Newman in *Community of Interest* (1980), for example, drew attention to the increased security and lower burglary rates of privatised streets in St Louis. In Britain, during the 1980s, the Adam Smith Institute, a right-wing think-tank, published a discussion paper, *Streets Ahead* (Elliot, 1989), advocating the privatisation of city and suburban streets as a solution to the problems of crime, environmental maintenance and local service delivery. Part of a wider libertarian project of advocating private solutions and provisions, the paper cited the existence of over 90,000 homeowner associations in the US, and over 30,000 neighbourhood crime protection patrols. It also noted that, at that time, over thirty million Americans lived in planned unit developments (PUDs) and master-planned communities (MCPs) often designed to be self-contained, defensible and exclusive. Many PUDs incorporate deed restrictions to ensure the provision of local services and conformity to the rules of the community association which often also defines eligibility for residence.

Segregated residential areas can be distinguished into two broad groups. The first, which we term 'enclaves', are where the means of segregation is relatively implicit. The second, which we term 'compounds', are where the means of segregation is explicit, active and physical. The two are not necessarily discrete and should be regarded as points along a continuum. Both are likely to consist almost exclusively of residential uses. Both involve some definition of territory: in the former, strangers are passively deterred from entering the territory; in the latter, strangers are actively deterred from entering the territory. In some instances, visitors need an invitation from a resident to enter the area.

Enclaves

'Enclaves' are where the means of segregation is relatively implicit (for example, by income or socio-economic group), passive and symbolic (for example, unmanned gates). Nottingham's Park Estate is a good example of an enclave. The Park Estate is a late Victorian middle class residential development on the former Royal hunting land adjacent to Nottingham's castle and the city centre. The south entrance has an electronic car control where only residents of The Park may enter. Pedestrians are unrestricted. The control is ostensibly to prevent the estate being used as a 'rat-run' rather than as a means of social seclusion. The other vehicular entrances have gates which are closed for one day a year to prevent the establishment of a 'right of way'. The premium attached to property prices has also given the population of The Park a relatively homogeneous socio-economic composition. Other examples of enclaves include Society Hill, Philadelphia; Beacon Hill, Boston; Nob Hill and the Castro, San Francisco.

Compounds

'Compounds' are where the means of segregation is explicit, active and physical (manned gates, walls, guards, private security guards). There are – as yet – few examples of compounds in Britain: the Bryant Mills complex in Bethnal Green – a former match factory converted into up-market apartments – is one example and there are others in the London Docklands. They are however very common in the US. Ellin (1996, p.70) for example notes that one in three new housing developments built in Orange County, the Palm Springs area and the San Fernando valley is gated. Furthermore, following the 1992 Los Angeles riot, hundreds of neighbourhoods applied to the City Council for permission to privatise streets and install or construct street closures causing local controversy. In addition to security, there are other motives for explicit segregation. Davis (1992a, p.14–15) notes that in Los Angeles some estate agents estimate that 'gatedness' raises home values by 40 per cent over ten years. He also notes how the residents of Whitley Heights were permitted to withdraw their streets from public use and erect eight electronic gates which restricted access to residents and approved visitors (Davis,1992b, p.173). One impact of 'gatehood' was a 20 per cent increase in local property values.

The sense of territoriality created by segregation is often further enhanced by the formation of Neighbourhood Associations, whose actions may range from organising neighbourhood watches to hiring security companies. The first Neighbourhood Watch in England was set up in 1982, and there are now 143,000 Neighbourhood Watch schemes in England and Wales covering six million households. Hope & Shaw (1988, p.12) define such schemes as 'the mobilisation of informal controls directed in the defence of communities against a perceived predatory threat from outside'. Ninety per cent of schemes see their role as being the 'eyes and ears' of the police (Home Office, 1996d). The assumption is that increased surveillance will deter criminals and, secondly, that increasing the contact between neighbours will lead to greater cohesion and solidarity in the face of the threat of crime and a consequent reduction in the fear of crime. Neighbourhood Watch schemes seem to work better where there is already a strong sense of community; as the Crime Concern's Chief Executive, Nigel Whiskin, observes, 'we need to put the emphasis on *neighbourhood* rather than watch' (1989, from Bottoms, 1990, p.14). Nevertheless, it is observed that Neighbourhood Watches are easiest to organise in middle-class neighbourhoods where crime rates are generally lower but more difficult to organise in poorer neighbourhoods which frequently have higher crime rates (Brantingham, 1989, p.342).

Acting collectively, residents may purchase further controls. Both passive and active means of surveillance can be reinforced by technology. CCTV cameras, for example, are becoming more common in residential areas, both on 'problem' estates and in up-market neighbourhoods. In London, for example, Belgravia and Hampstead Garden Suburb have recently set up privately-operated CCTV systems. Where surveillance is not sufficient and the intervention of a capable guardian is required, then as Davis (1990, p.223) describes: 'The carefully manicured lawns of Los Angeles's Westside sprout forests of ominous little signs warning: "Armed Response!".'

Containment

The counterpoint of segregation – and perhaps the ultimate extrapolation of the desire for segregation – is 'containment'. Rather than excluding non-residents from certain areas, the residents of certain areas may be 'contained' within their own areas. Some cultural, ethnic, religious or social groups may band together in the interests of solidarity, but their 'ghettos' result from the exercise of choice. By contrast containment is an enforced ghetto-isation. Charles Murray's concept of 'drug-free zones for the majority' 'may require social refuse heaps for the criminalised minority' (Murray, 1990a, p.25). As Davis questions, if this means concentrating the bad apples into a few 'hyper-violent, anti-social neighbourhoods', how will the underclass be confined to its own neighbourhoods and kept out of the 'drug-free shangri-las of the over-class'? Technology can come to the rescue and Davis (1992a, p.12) is able to answer his own question:

> Drug offenders and gang members can be 'bar-coded' and paroled to the omniscient scrutiny of a satellite that will track their 24-hour itineraries and automatically sound an alarm if they stray outside the borders of their surveillance district.

Integrated residential areas

In contrast to segregated residential areas, 'open' residential areas are wholly integrated into – and undistinguished from – the surrounding city. Rather than being purely mono-functional areas, they will also tend to have a greater or lesser element of mixed-uses. As public access must be retained due to the mixed use, non-residents and 'strangers' cannot be excluded. Thus, techniques of explicit segregation are less appropriate in central cities than in suburban and inner-city settings. As the impact is less detrimental to the general well-being and safety of the city centre as a whole, exclusion at the block level, however, is more acceptable than at the area level. The Capitol Hill area in Seattle is a good example of an integrated neighbourhood. As there are no differentiating physical characteristics in terms of housing types, it is difficult to define its boundaries precisely. It is also a socially heterogeneous community where different age, social and sexual groups reside. In many ways it has the rich urban life and natural surveillance that Jacobs advocated.

Hillier's principal criticism of Newman's and Coleman's ideas is that they advocate defensible enclaves which exclude the natural movement of people; as a result of implied local social controls, 'strangers' are unable or unwilling to enter the space. Hillier (1988, p.86) does accept that there is a stronger feeling of territoriality in segregated spaces and people are more likely to challenge the presence of strangers, he argues that this is (only) because people felt intruded upon and more unsafe. However, he argues that the treatment of defensible space by Newman and Coleman tends to assume that all strangers are dangerous. His argument is that the presence of people is the primary means by which space is naturally policed. The more this is eliminated, then the greater the danger created once a potential criminal has appeared on the scene. He, therefore, concurs with Jacobs' view that feeling safe in a city depends largely on areas being in continuous occupation and use. It is also significant that Hillier is discussing both property and personal safety and that personal safety includes that of residents and non-residents while the other writers are principally – although not exclusively – concerned with property safety.

Hillier developed his ideas in his concept of space syntax (Hillier *et al.*, 1978; Hillier & Hanson, 1984). In this concept, Hillier argues that there are certain desirable spatial characteristics which increase the likely presence of people and thereby enhance the feelings of safety. Rather than being enclosed and self-contained, spaces ought to be opened up and integrated with other spaces, so that the pedestrian is encouraged to see into and move through them. The concept of permeable, integrated environments, however, gives greater importance to spatial patterns and layout than to land use in determining pedestrian use. This emphasis contrasts with Jacobs' argument for mixed land uses to provide safer environments.

SAFER URBAN HOUSING

At the block or street level, whether in segregated or integrated developments, a fundamental concern of those who desire to live in the central areas of cities is that the housing is sufficiently safe and secure. It must also have safer parking and the residents must feel sufficiently confident to walk the streets and enjoy the local urban culture which is likely to be one of the primary attractions of the location. Wekerle & Whitzman (1995, p.109) argue that safe residential areas are perhaps the most important element of safe cities. While there is often a choice involved in the avoidance of an unsafe downtown or unsafe public transport for many people there is little choice in where to live. Nevertheless, for many women the home itself is not a safe haven. As shown by the British Crime Survey (Mayhew *et al.*, 1993), the most numerous type of assault was incidents of 'domestic violence', eight out of ten of which were against women. Nearly half of the assaults mentioned by women in the survey were of this type.

In terms of the design of safer urban housing, it is useful to note the subtle safety features of successful precedents of urban housing. The traditional terraced housing and streets of town houses are examples of safer housing because they deploy the situational crime prevention techniques of access control, entry/exit screening and surveillance. The twitching net curtain, for example, provides discreet natural surveillance. Housing like this in Philadelphia around Rittenhouse Square provided the basis for many of Jacobs' observations. The key characteristic is a distinct difference between the public 'front' and private 'back'. At the front, various design mechanisms place residents in control of the street immediately in front of their house: a short flight of steps up to the front door; gates, walls and/or front gardens create distance and symbolic barriers; the ground floor of the house is often above the level of the street, while projecting bay windows permit surveillance up and down the street and, crucially, of callers at the front door. In their attempts to enhance privacy, many Modernist housing designs shut themselves off from the street, thereby preventing natural surveillance of the street. There are many examples of streets of modern housing where few windows face onto the streets, rendering the streets less safe (**Figure 10.1**). In such streets, it is virtually impossible to create effective Neighbourhood Watch schemes, as few dwellings have a clear sight of the street.

The most common housing in city centres, however, is multi-occupation blocks of flats and apartments. Where there is an entrance serving a number of dwellings, there should be some form of entry control – perhaps with a concierge – and the number of dwellings served by that entrance should be severely limited; the more doorways onto the street, the busier the street will be. This is in accordance with the tenets of the situational approach, the effect of which is to lessen the residents' fear of crime by placing them in a position of control. The private back should be exclusively for residents, consisting entirely of individual private gardens or a communal space and protected by lockable

Figure 10.1 Housing in Society Hill, Philadelphia. The design of the housing from two different periods illustrates the issue of providing 'eyes onto the street'.

gates. The shared private space is protected by the mutual surveillance of residents. Again this is in accordance with the tenets of the situational approach and, once more, the effect is to place residents in a position of control. Residents also require safe parking. If their cars cannot be garaged, they must be in close proximity and in direct view of the dwelling.

The least attractive urban housing is expressly 'target hardened'. The image of such housing is generally negative giving the impression that it is a highly-precarious location in which to live. Wekerle & Whitzman (1995, p.109) list the:

> locked high fences, intercom entry systems, security patrols, and video surveillance [which] have become elements of some middle – and high – income residential complexes. The security offered by these measures, however, can only be seen as illusory: voluntary incarceration in a kind of prison . . . in order to avoid violence.

ENCOURAGING OPPORTUNITIES FOR CITY CENTRE LIVING

In the UK, there has been official encouragement for more residential development in town and city centres. Recent planning guidance (DoE, 1996, section 2.13) has advocated enhancing the vitality and viability of city centres by increasing the number of people living there:

> Local planning authorities should use their development plans to set out policies for mixed-use development in town centres by identifying suitable areas and sites, and issuing planning briefs. A mixture of small businesses, houses or offices in or near town centres and the occupation of flats above shops, can increase activity and therefore personal safety, while ensuring that buildings are kept in good repair. Residents and workers stimulate shopping, restaurants and cafes, and other businesses to serve them, and so in turn add to vitality.

Nevertheless, despite a generally favourable planning policy context, relatively little residential development has occurred in city centres. A prime factor inhibiting potential developers or conversion projects is that there is still little demand for housing in such locations. Demand did, however, significantly pick up in the 1980s with the 'yuppy' lifestyle becoming fashionable. Rather than a lack of interest, a key factor preventing people moving back into city centres may be the lack of economic opportunity – the quality of housing at a price that they are both willing and able to pay – and the physical opportunity. To address these issues, there might be a case for demonstration projects with a sufficient level of public subsidy to 'test the water'. Their success or failure will demonstrate whether a real demand exists, although developers might still remain cautious and uncertain of the extent and depth of the demand.

The principal development opportunities in city centres are new build and/or conversions. Apart from smaller infill sites, new build sites are few and far between in city centre locations and may frequently be appropriated for a higher value use. There are also residential opportunities through the conversion of buildings either entirely or partially, for example, through 'living over the shop' schemes or the conversion of upper floor(s) of a building. With the increasing decentralisation of many activities and the development of high quality new office space, there is surplus office and industrial space that may become derelict and blight the city centre. Conversion to a different use is a way of ensuring that these buildings remain in active use. The capacity for conversion is, however, limited by a number of factors: the physical and spatial parameters of the existing building; the architectural character of the building and the constraints placed by special historic building controls on change; the planning policy context; the environmental consequences of the change of use, particularly in terms of traffic generation and management; and the reception of the commercial market and possible users and investors to the change of use. Nevertheless, the residential function is relatively flexible and can be adapted to fit many forms.

Living Over the Shop

One of the major opportunities for more dwelling units in city centres is through the conversion of the vacant space above shops. Every city centre in Britain has a significant amount of vacant or underused space above its shops. In addition to increasing the resident population of town centres and city centres, bringing greater vitality to these areas and improving the safety of the city centre, especially at night, further benefits accrue from the development of this space into flats, such as the physical improvement of a building's frontage and of the street where the upper floors appeared vacant or derelict. It also provides surveillance of the retail properties outside shop hours. There is general policy support for such conversions in many of the recent PPGs, but local authorities seem unable to secure the implementation of these policies and the development of these schemes. In the UK, a commitment to utilising the space above shops began in 1989 with the setting up of the Living Over The Shop Project by Ann Petherick (see Petherick & Fraser, 1992).[1] A number of local authorities and some large property owners – notably the Nat West Bank and Boots The Chemists – set up their own initiatives. The DoE also set up a pilot scheme – Flats Over Shops (FOS) – between 1992–95. Nevertheless, results have tended to be disappointing and much of the potential remains unrealised.

Further progress, in particular, needs to be made in approaching national property owners who own the vast majority of retail premises. Their involvement could significantly increase the number of flats over shops. Part of the

1 The approach's central features are the involvement of a housing association and the use of a commercial lease. The freeholder of the property grants a fixed-term commercial lease to a housing association for the upper-floors of the property who then grant an Assured Shorthold Tenancy (AST) to residents and manage the tenancy. The combination of the AST and the commercial lease ensures that the property owner has vacant possession on or before the expiry date of the commercial lease, while the involvement of the housing association enables access to Housing Corporation funds for conversion and renovation.

problem in encouraging living over the shop schemes, and mixed use developments more generally, is the entrenched attitude of funding institutions which extends not just to an aversion to mixed retail and office developments but also to any other land use that may involve more than minimal management (Punter, 1990, p.11). In the US, Duany & Plater-Zyberk's Traditional Neighbourhood Development (TND) and Peter Calthorpe's Transit-Oriented Development (TOD) both advocate traditional living over the shop type arrangements through building design and design codes (see Katz, 1994).

A more recent development emerging since the property crash of the late 1980s is the conversion of redundant offices to residential use. Most cities have an abundance of unlet office accommodation, especially in buildings dating from the 1960s. These office buildings are no longer suitable for modern office needs. As a consequence, many such buildings, especially in London, are being converted into multi-occupancy blocks with as many as 70–100 flats. Such conversions have the potential to bring in residents, underpin mixed-use and help people the streets in the evenings. The Oxo Tower Wharf building on the south bank of the River Thames is one of the most mixed-use conversions. The project includes shops, restaurants and bars, workshops, a museum and co-operative housing. Achieving this degree of mixed use has not been easy and is a testimony to the community-based owners, Coin Street Community Builders, being unwilling to compromise their vision of a mixed-use development (Coupland, 1995).

Urban Villages and Urban Quarters

City centre living should generally be encouraged on an area basis, with measures to improve the locality planned on a comprehensive rather than piecemeal basis. The conversion of individual buildings or one-off residential developments may make little contribution to the safety of an area. Punter (1990, p.14) describes isolated residential developments which fail to link to another and therefore do not provide 'either a continuity of community, a network of surveilled routes, or the critical mass necessary to engender a sense of security to support complementary activities'.

Planning for residential development on a more comprehensive basis usually requires public authorities to provide infrastructure that is supportive of residential development, such as the provision of car parks for residents to remove the necessity of on-site provision for each individual development. Equally, parts of the city centre can be promoted and developed as residential quarters with planning authorities providing positive support through: marketing sites; feasibility studies; development briefs; and 'talking up' development and building confidence in the city centre as a potential residential area.

A key idea in this context is the promotional value of the concepts of 'urban villages' and 'urban quarters' (see Urban Villages Group, 1992; Carmona, 1996). Creating a vibrant 'urban village' in an area without a tradition of residential population, however, needs a critical mass of development and inevitably takes a number of years to occur. As many central areas do not yet have an infrastructure that would support residential uses, the initial residents are inevitably pioneers. Often the image of an area needs to be improved and purposefully reconstructed to suggest that this is a desirable – perhaps 'alternative' – residential location. Environmental improvements might be crucially important in changing the image and perceptions of the area. Other public sector actions could also make city centre development more attractive by restricting the competitive supply of housing opportunities.

Nascent new residential quarters or urban villages are emerging close to many British city centres. At the area level, such developments tend to be open integrated developments in physical terms. An element of segregation is, however, introduced by virtue of their selling price and the ability to afford to live there. Many of these have arisen for conservation reasons as a way of using old buildings. A good example is Glasgow's Merchant City where a planned process led by the local authority and involving considerable public subsidies led to the functional restructuring of the quarter, provided a use for historic buildings and, through the creation of a viable residential market in an area with no tradition of such uses, brought people to live in and near the city centre.

By the early 1980s, the rundown appearance of Glasgow's historic Merchant City was blighting the city centre and, more generally, inhibiting other initiatives promoting Glasgow, such as the Glasgow's Miles Better campaign. As the local authority had acquired a major landholding in the area, it was a key party in leading the revitalisation process and decided on a housing-led revitalisation. The first demonstration project was the Albion Building which was converted in 1982 into twenty-three flats aided by a significant public subsidy. A number of developments followed based on similar financial arrangements: conversion grants plus cheap transfers of property.

The most significant project in the Merchant City was the Ingram Square development. Although when initiated, a market for housing in the Merchant City was becoming more established, development costs outweighed estimated sale value. Public sector gap funding support was still necessary but the scale of the project (240 flats) put it beyond the reach of the *ad hoc* partnerships previously devised. Instead of the grant plus property arrangement, a joint development company was formed involving the district council and the public sector Scottish Development Agency (SDA) as part investors in the project with the private developers (CGDC, 1992, p.3).

The housing-led revitalisation of the Merchant City has been successful. The district council's efforts in promoting and supporting conversion work with public subsidies, turned the liability of owning dilapidated and often empty buildings into the positive advantage of being able to control the pace and quality of development and to achieve the conservation and revitalisation of an historic quarter of importance to Glasgow. Since the first conversion in 1982, more than 1200 flats have been created providing a substantial indigenous population and a demand for other facilities and contributing to the vitality and safety of the city centre generally and the Merchant City in particular (**Figure 10.2**).

Figure 10.2 The Italian House in Glasgow's Merchant City. Residential developments and conversions in the area together with environmental improvements have helped create vitality and life in the quarter's streets.

CONCLUSION

This chapter has argued that residential developments in city centres will increase the peopling of the city's streets. It is this increased peopling that would make the city centre safer. Given this argument, the key challenge is to encourage further residential development in city centres and to broaden the range of people who have access to this type of housing. Until the suburbanisation of UK and USA cities, most central cities included residential areas and even in the commercial cores, living over the shop was common. The middle class flight to the suburbs, encouraged and supplemented by Modernist zoning policies, robbed the city centres of their complexity and vitality, as well as their safety. Even at the height of residential decline, however, pockets of residential areas remained. Rittenhouse Square and Society Hill in Philadelphia; Beacon Hill in Boston; Nob Hill in San Francisco; the Upper Eastside in New York; Belgravia, Knightsbridge and Chelsea in London; and the New Town in Edinburgh are notable examples. What these areas have in common is that they are inhabited by high income residents able to buy security or, at least, influence policing priorities in their areas.

What has also been seen since the 1980s is a renaissance of city centre living as increasing numbers of the active retired, childless couples and people with alternative lifestyles choose city centre living. This increase in numbers has resulted in increasingly peopled streets. City centre living may arise in non-traditional residential locations, such as historic warehouse districts, where the first residents are pioneers. Closely following the pioneering of central city living, the broadening of that foothold is often by those in higher income groups. For city centres to have a greater residential population, however, the catchment must extend beyond narrowly elite groups. As well as physical access, there must also be economic access; established residential enclaves in the central city are often exclusionary by dint of incomes as well as by physical means.

The prerequisite for city centre living is that it is safe. In a context where housing provision will be predominantly by the private sector, the housing produced will need to appeal to the private developers' customers. As there remain concerns about safety both on the streets and of property, there are gated communities, segregated residential areas, enclaves and compounds. The general contribution of such areas to city centre and central area safety is mixed. Although increasing the city centre population is beneficial, segregated residential areas do not increase the area or terrain of the city centre which becomes peopled and arguably safer. There is security within the compound or enclave, but not necessarily outside it. The principal criticism, however, of segregated residential areas is that they are exclusionary. Fainstein (1994, p.235–6) neatly summarises this problematic issue:

> The desire of people to live with others sharing similar outlooks and modes of behaviour is understandable . . . the wish to live in personal safety is obviously legitimate . . . and is most easily obtained through the creation of boundaries.

She further argues that it may be considered acceptable for subordinate groups to separate themselves if they so choose. Although by making this choice they are not usually denying other groups a material benefit, it nevertheless complicates the issue by undercutting the principle of social integration. In this respect, socially inclusive activity corridors are preferable to socially exclusive private or privatised 'enclaves' or 'compounds'.

In practical terms, it is often the *degree* rather than the *fact* of exclusion that is at issue. As Punter (1990, p.13–14) notes, there is often a fine balance between the concern for integration and people's desire for segregation and between 'the kind of positive policies which advocate greater surveillance and demarcation of the public realm . . . and those reactionary impulses which aim to exclude all outsiders in a bid to secure exclusivity and to protect land values'. Privatisation and segregation diminish *both* the amount of genuinely public space and the number of people to provide the informal social controls to police the public realm. Thus, what is more preferable for the safety of the city centre as a whole are residential areas which have a greater degree of integration, make a positive contribution to the peopling of the city's streets and create more eyes on the street.

11

The Twenty-Four Hour City Concept

The Twenty-Four Hour City concept is a new approach to revitalising and creating safer city centres. In the UK, its origins can be traced to 1993 when Manchester launched its Olympic bid and ran the More Hours in the Day initiative. Since then the idea has spread rapidly and the approach – or elements of it – has been adopted by most British cities. The concept is influenced both by those cities in continental Europe which are inherently twenty-four hour in their nature and those which, since the 1970s, have developed cultural policies to revitalise their urban nightlife (Bianchini, 1995). Indeed, an explicit aim of many British cities is to become a 'European' city.

Rather than as just nine-to-five retailing and office centres, the Twenty-Four Hour City concept provokes a more holistic and expansive consideration of the totality of city centres, including their social and cultural dimensions. It also has an aspirational dimension aiming to stimulate ideas and entrepreneurial activity. A review of British initiatives (Stickland, 1996) reveals the concept's three primary elements:

- providing a safer city centre for a broader spectrum of the population to enjoy without fear;
- developing the evening economy of the city;
- improving the city's image to attract inward investment.

The concept is based on three presumptions: first, if people are involved in a diverse range of leisure activities beyond the traditional hours of nine-to-five, the peopling of the city centre increases and the level of natural surveillance improves. Furthermore, the provision of alternative attractions to the traditional pub and club-based culture, could encourage users who, at present, do not use the city centre at night due to fear of crime. A more diverse cross-section of the population using the city centre at night would have a multiplier effect on the local economy. The city centre is also likely to become a safer and more accessible place. Secondly, changing employment patterns combined with increasing disposable income, provide the potential for places to change from production-based to consumption-based economies, thereby generating

Copyright © 1997 Tim Heath and Robert Stickland

employment opportunities in the predominantly evening-based leisure industries. Thirdly, a vibrant city centre at night is regarded as a positive attraction to inward investment. Visitors generally stay in centrally located hotels, eat in restaurants and may look for evening entertainment, especially if these areas are safe. Thus, if their experience of a place is positive, it is anticipated that they will return, whether as a tourist or a corporate investor.

THE TWENTY-FOUR HOUR CITY

The concept of twenty-four hour cities is not new; indeed, many cities, such as Edinburgh and many in continental Europe already have twenty-four hour characteristics. The concept of a more-planned night-time economy has been established in Europe since the late 1970s. Bianchini (1995, p.121) identifies Rome's programme of night-time cultural events as being one of the first such initiatives. In 1977 Renato Nicolini, the Rome City Councillor in charge of cultural policy, started an annual summer cultural programme:

> focused on public monuments and historic streets and squares in the city centre, which were made safer, more attractive, and more accessible by coordinating cultural policy with appropriate policies on public transport, lighting, policing, childcare, and environmental improvements (Bianchini, 1990, p.7).

Cultural policies of this nature have been a common part of the night-time vitality of a number of European cities. What is new, however, is the commitment in Britain to twenty-four hour strategies and initiatives in order to achieve city centre revitalisation and to create a safer environment.

The 1980s saw the re-emergence of concern for city centres and urbanity. O'Connor (1993) identifies the factors accounting for this:

- The de-industrialisation of older industrial areas resulting in derelict land in city centres.
- The development boom of the 1980s and the associated rise in centrally located offices reclaiming some town centre prestige.
- The rise in competition between cities at both national and international level leading to the increased importance of city image.
- The reorganisation of city centres around consumption, rather than production, and giving rise to a new landscape of buildings. This has, however, often resulted in place-sameness and city centres losing their sense of place.
- The shift from production to consumption that had previously been considered marginal to the primary function of cities, thereby creating opportunities for cultural creation and entrepreneurial activity.
- Since the 1960s, cultural definitions have changed under the pressure of transformed lifestyles and 'commodification'.

Changes in social aspirations and the changing ratio of work to leisure time are explored in Comedia's *Out of Hours* report (1991). The report describes how the growth of flexible working hours, shift and part-time work, and even the video recorder, have given rise to a greater choice of evening entertainment and a different temporal demand for leisure activities. The report argues that; 'these shifts in people's lifestyles and needs must be recognised in town centre strategies' (Comedia, 1991, p.22). The study identifies a number of problems contributing to the lack of night-time activity in Britain's city centres, including:

- the lack of things to do at night-time
- the existence of a pub-culture and its lack of interaction with the street
- the monofunctionality of city centres – office and retail focus
- pedestrian-unfriendly environments
- poor public transport provision at night-time
- restrictive licensing laws
- low consumer disposable income
- single activity visits of city centre users.

Comedia also highlighted the problem of unsafe city centres, with the perception – particularly among women, the elderly and middle-class inhabitants – of youth domination and crime. The report recommends that a holistic view should be adopted to address the problems of fear of crime and the exclusion of many ethnic and age groups from the city centre. Initiatives such as the Reclaim the Night women's movement have sought to raise consciousness of this safety and perception issue.

The reduction of perceived and actual night time crime is a central aim of many Twenty-Four Hour City schemes. As the DoE Circular, *Planning Out Crime* states:

> One of the main reasons people give for shunning town centres at night is the fear about their security and safety: one of the main reasons for that fear is the fact that there are very few people about. Breaking that vicious circle is a key to bringing life back to town centres. By adopting plan policies that encourage a wide and varied range of uses, local authorities foster the creation of lively, attractive and welcoming environments. This objective may require wider measure . . . They may well extend, for instance, to enabling arrangements that help to promote the night economy. (D0E, 1994, para 14, pp 3)

This statement also supports the key issue – critical to the success of the Twenty-Four Hour City – of creating demand for the city centre at night.

The Role of Culture

Discussing the social changes that have taken places over the last twenty years, Lovatt (1993) argues that until recently the regulatory and development strategies deployed at both national and local level missed the potential and opportunities afforded by the increasing demand for leisure and recreational activities. O'Connor (1993) also identifies the link between culture and the night-time economy, noting that night-time activity is implicit with these functions, yet cultural policy in the UK has, unlike continental cities, marginalised this part of the day. Rather than developing its potential, the overriding objective was to control and regulate the night-time economy through safety, planning, licensing and health policy. Lovatt (1993) concludes that the emergence of a cultural infrastructure and the increasing value of city centre sites during the late 1980s occurred *despite* rather than *because* of the institutional, central and local government development strategies.

Identifying the need for new and innovative ways to solve city centre problems, Landrey & Bianchini (1995, p.10) argue that approaches should capitalise upon a city's cultural and social potential. They call for the 'creative energies' – such as carnivals and festivals – to become part of mainstream city life. This could be achieved through the promotion of theatres, evening classes, conferences, cafés and clubs, as well as street animation and activity in the night-time economy. The Twenty-Four Hour City approach provides a holistic view to these issues; Comedia (1991, p.51), for example, advocated that:

> councils should change their attitude towards the use of their town centre by regarding it less as a nine-to-five retailing and employment centre and more as an eighteen hours a day, seven days a week, economic, social and culture centre.

The Civic Trust's (1993) *Liveable Towns and Cities* highlighted the importance of cultural initiatives and niche markets as well as the revitalisation of local public social life, civic pride and identity. Although cultural industries are closely linked to cultural animation and festivals, their role in Twenty-Four Hour City initiatives encourages a more diverse range of users and improves city image, as well as providing people-policing and natural surveillance. The Civic Trust (1993, p.75) also argue that cultural animation can help to raise levels of participation in public life for a diverse range of social groups, particularly as these can often incorporate the 'skills, and talents of young people, women, ethnic minorities or the unemployed'.

By the late 1980s, local authorities and others involved in cultural industries had recognised the changing role of culture and had began to formulate cultural strategies aimed at revitalisation. Recent strategies for the revitalisation of city centres have, therefore, increasingly supported the development of the evening economy through the use of creative and cultural strategies. The strategies are promoted on their ability to capitalise upon changing labour, economic and social conditions that create a new opportunity for the city centre as a place to capture the increasingly mobile consumption capital. As a result, planners, policy-makers and town centre managers are now talking to a new set of players in the city centre at night – not as controlling authorities – but as partners in the revitalisation of the city centre. The initiatives include the promotion of festivals and other cultural activities, central office and residential developments; all of which incorporate aspects of both social and cultural vibrancy.

TWENTY-FOUR HOUR CITIES IN BRITAIN

Since the late 1980s, the Twenty-Four Hour City approach has been advocated by a number of consultancies and other institutions, including Urban Cultures

Table 11.1 Reasons for adopting a twenty-four hour city strategy (source: Stickland, 1996).

Rank	Reason	%	Authority
1	Safer city	91%	Bristol, Cardiff, Glasgow, Hackney, Leeds, Liverpool, Manchester, Newcastle, Nottingham, Sheffield
2	City image/ inward investment	64%	Bristol, Cardiff, Glasgow, Leeds, Liverpool, Manchester, Nottingham
3	Economic regeneration	54%	Bristol, Cardiff, Hackney, Manchester, Newcastle, Sheffield
4	Local service provision	36%	Cardiff, Hackney, Manchester, Sheffield
5	Specific local reason or event	27%	Manchester, Newcastle, Sheffield
6=	Following another city's lead	18%	Islington, Newcastle
6=	Prompting by the DoE	18%	Islington, Sheffield

Ltd, Urban and Economic Development Group (URBED), and the Manchester Institute of Popular Culture. Initial interest in the approach was generated by Manchester City Council's attempts to recapture its evening economy, culminating in the Twenty-Four Hour City: The First National Conference on the Night-time Economy in September 1993. The approach gathered momentum with Leeds and Cardiff developing their own approaches and initiating structures and programmes. Many cities, including Newcastle, Liverpool, Islington, Hackney, Sheffield, Nottingham, Glasgow, Camden and Bristol have all developed night-time initiatives. The approach is now being adopted by most large urban areas with established city centres. Nevertheless, despite having distinct centres and large populations, it has been less widely adopted by Metropolitan Districts and London Boroughs.

A survey of local authorities (Stickland, 1996) showed their reasons for implementing the Twenty-Four Hour City concept (**see Table 11.1**). The principal reason was its value as a means to improve the night-time safety of their city centres. This is supplemented by a strong desire to attract inward investment and to act as a catalyst for economic regeneration.

The operational hours of the schemes vary between cities. There are two groups that might be attracted to the city centre after office hours and on into the night. First, those who are shopping late, and possibly visiting a restaurant or café, and who then leave the city centre. Secondly, those who arrive later in the evening, with the purpose of consuming entertainment, whether it be an art gallery, gym, cinema or pub. Cities adopting Twenty-Four Hour City schemes can, however, be divided into three groups in terms of the timescale of their schemes. First, those – such as Hackney and Glasgow – which are concerned predominantly with an evening economy that bridges the gap between office closure and the pub and restaurant activity. Secondly, those interested in developing a night-time as well as evening economy such as Bristol, Sheffield and

Liverpool, and finally those interested in a full twenty-four hour city. Only Cardiff claims to have a fully operational Twenty-Four Hour City, with activity operating throughout the day. Manchester and Leeds both state that it is their aim to develop a fully functional Twenty-Four Hour City.

Manchester, Cardiff and Leeds are the three cities that have taken the lead in promoting the Twenty-Four Hour City approach. Manchester and Leeds have integrated initiatives in the majority of their strategic publications, including the Manchester Unitary Development Plan (UDP) and the draft version of the Leeds UDP. These cities have also adopted the twenty-four hour theme with varying aims. Manchester has a broad set of aims, centred around economic regeneration. Leeds is concerned with its promotion as a European Centre for inward investment. Cardiff balances these two aims, looking principally to its promotion as a capital city.

Manchester

Manchester, a city of over two and a half million people in north-west England, has adopted a broad approach to its Twenty-Four Hour City strategy and is the only city to identify the 'recapturing' of a previous twenty-four hour city as one of its aims. Manchester considered that the erosion of what had been a twenty-four hour city centre was due to economic and social changes since the 1960s. In particular, as contributory factors to the decline of its city centre at night, the council identified the relocation of the print works and wholesale fruit and vegetable market from the city centre and the decline of cafés and other leisure activities due to private entertainment. The city council attributes many of the associated night-time city centre problems as a being inextricably linked to this loss of activity. In its *City Development Guide*, the council also acknowledges the role that past planning policies have had in the decline of the city centre:

> Past planning principles which emphasised rigid zoning of activities and housing tenures have proved unhelpful because they created areas which are dead at certain times of day. (MCC, 1995, p.5).

The opportunities to recapture its past vitality were emphasised by specific events that demonstrated the demand for an improved city image and provision of night-time service activities. These included the development of the third international terminal at Manchester Airport, the Olympic 2000 bid and the city's status as the City of Drama in 1993. These were all cited in the Council's May and June 1993 reports to the city centre sub-committee of the Policy & Resources Committee as being contributory factors in initiating the Twenty-Four Hour Pilot Scheme which took place in September 1993. In addition, the council cites other reasons, such as: city safety, economic development of the potential of an evening economy, and the need to develop as a regional centre and as a European city. These aims have been adopted and included throughout the new Manchester UDP, not necessarily as specific twenty-four hour policies but as a consistent theme. As such, Manchester has the most developed and integrated approach of any British city.

Leeds

Leeds is the second largest metropolitan district in England with a population of 700,000. The city has adopted a more outward-looking approach focusing on the city's role as a European City. Its two main aims are to attract inward investment and to create a positive city image. Although this is also one of Manchester's aims, in Manchester it is a broader goal with a more balanced approach between inward-looking attempts to enhance the city centre for its own population and to capture mobile capital investment. The main emphasis of Leeds' approach is to promote a positive city image as a European centre. This is reflected through the lead agency, the Press & Public Relations Department, whose aim is capturing inward investment in a European market.

Manchester and Leeds have adopted contrasting management structures for their initiatives. While in Manchester there is joint structure involving the Planning Department and the city centre management team, in Leeds the departmental responsibility lies outside of the Planning Department. Although all of the local authority departments are required to consider the city centre at night, a special unit within the Press & Public Relations Department – the Twenty-Four Hour City Team – was created to manage the implementation of the various projects, which indicates the strength of its commitment to the approach.

Cardiff

Cardiff – the capital of Wales with a 280,000 population – has combined Manchester's emphasis on positive improvements to its city centre, with Leeds' city image emphasis. Cardiff's approach is centred around its promotion as a capital city, and as a regional and European centre. This aim is being tackled through the development of a safer environment, the economic development of the evening economy and the provision of local services for the consumption of its own inhabitants. The approach in Cardiff has also gone through an evolutionary process in reaching its objective of becoming a true twenty-four hour city. Cardiff has adopted a less formal structure for the delivery of a Twenty-Four Hour City consisting of a series of initiatives within different departments. In terms of cross-departmental co-operation, it is, however, moving quickly to the stage reached by Manchester and Leeds (Stickland, 1996). The recent *Cardiff City Centre: Prospectus for Investment* (CCC, 1995, p.5) states that one of the strategic objectives for the city centre is:

> To encourage the establishment of an accessible '24 hour city' to underpin and widen the cultural and commercial offer and signify the city centre as an evening destination for visitors and residents within the conurbation.

TWENTY-FOUR HOUR INITIATIVES

Local authorities have utilised a variety of initiatives in developing Twenty-Four Hour City schemes (**see Table 11.2**). Most of these initiatives assist in fulfilling the following objectives: economic development; making a safer city

Table 11.2 Twenty-four hour city initiatives adopted by local authorities (source: Stickland, 1996).

CITY	TIMESCALE	KEY INITIATIVES
Leeds	24 hours	Licensing, Retail, Cafes/Rest., Lighting, CCTV, Festivals, Theatre.
Manchester	24 hours	Licensing, Retail, Cafes/Rest., Lighting, CCTV, Festivals, Street Events.
Sheffield	20 hours	Licensing, Cafes/Rest., CCTV, Cultural Ind., Theatre.
Liverpool	20 hours	Licensing, Cafes/Rest., CCTV, Theatre, Street Events.
Cardiff	24 hours	Cafes/Rest., Lighting, CCTV, Festivals.
Glasgow	18 hours	Retail, Cafes/Rest., Lighting, CCTV.
Nottingham	20 hours	Retail, Cafes/Rest., Theatre, Street Events.
Newcastle	20 hours	Licensing, CCTV.

centre; and improving the city image. Not surprisingly, Manchester and Leeds have incorporated the broadest range of initiatives. Cardiff, however, has surprisingly few initiatives, with a strategy concentrating on a smaller range of policy tools. The initiative most commonly used is the installation of CCTV, closely followed by the promotion of cafés and restaurants and strategies aimed at tackling the issue of licensing.

Licensing Issues

The idea of using licensing initiatives to create safer city centres preceded the Twenty-Four Hour City concept. Indeed, Hope (1985, p.57) proposed using licensing controls to ameliorate drink-related disorder in city centres and defined two types of disorderliness affecting the city together with possible remedies. He argued that time-specific disorderliness could be ameliorated by: staggering pub closing times avoiding concentrations of people; increasing the number of late-night licences through permissive licensing; and/or facilitating the dispersal of people from the city centre. *Situation-specific* disorderliness could be ameliorated by: altering the number and density of licensed premises; discouraging the concentration of youth-oriented leisure facilities in city centres; and/or reducing the amount of indefensible public space in city centres. Many of the Twenty-Four Hour City initiatives incorporate Hope's ideas.

The licensing issue is a central part of many Twenty-Four Hour City schemes and most authorities have tackled the issue in some form or other. The most advanced of these are Manchester and Leeds' attempts to influence the licensing laws. The licensing debate operates at two levels: first, national attempts to alter the present system of licensing restrictions and the issuing of licences, and secondly, attempts to change local restrictions on licensed premises. This initiative has a significant safety dimension. In most cities uniform closing times often mean an increased amount of incidents at closing time, which discourage many potential users from using city centre facilities. Manchester and Leeds are tackling the first issue, while Newcastle and Sheffield are tackling the second.

Manchester and Leeds have made the most concerted efforts to bring about a review of licensing laws. In June 1993, Manchester issued its response to the Home Office consultation paper, *Possible Reforms To The Liquor Licensing Laws in England And Wales*. Manchester argued that the absolute discretion of granting liquor licences held by magistrates should be abolished and that statutory powers should be established. Manchester is also lobbying for a change in the issuing of licences, to bring it in line with the Scottish system. This allows local authorities the power to issue and determine the operating length of licences similar to the system for issuing Public Entertainment Licences. Manchester's lobbying of central government has been supported by Leeds. Leeds has run seminars and formulated *Leeds City Council – 24 Hour Licensing* (LCC, 1995) as a policy statement. Nevertheless, little can be achieved until the general law on licensing hours, requiring a special licence for extensions beyond 2 am, is removed. By staggering closing times, both Leeds and Manchester have attempted to reduce the number of people on the streets and the strain on public transport systems at peak periods. They have also attempted to cater for increasing numbers of people over a longer period of time by extending the hours of public entertainment licences.

One of the key initiatives of Manchester's pilot study, More Hours in the Day Campaign in September 1993, was an extension of permitted opening hours for pubs and bars until midnight and for clubs until 4 am. In March 1994, Manchester City Council issued its *Report For Resolution* reviewing the initiative, for the Policy & Resources Committee. It reported that more people had visited the City Centre during this period and that: 'the police had to deal with fewer public order and street crimes related to alcohol consumption' (Manchester City Council, 1994a, p.71). During the festival, Greater Manchester Police carried out a survey of crime which indicated that crime in general fell by 43% and drink-related incidents fell by 16% (from Stickland, 1996); furthermore, taxi firms reported 'increased trade and less aggravation' (Lovatt, 1994). The improvement was achieved despite increased bar and door takings in participating establishments. The evidence from Manchester would therefore suggest that the controlled relaxation of licensing hours can have a positive impact upon city centre crime helping to reduce personal and property crime.

The Newcastle initiative involved extending pub opening hours to 1 am and club hours to 3 am during the 1996 European Football Championships. In Sheffield, the City Council campaigned to end a fifteen year period during which no new club licences were issued in the city. The main obstacle was the Magistrates Committee which issued the licences. The initiative has proved successful and night-time activity has increased with the granting of more licences. Initiatives such as those in Sheffield and Newcastle, however, must still operate within the national legislation for operating hours. A change in the law will be required if licensing initiatives are to be achieved in other cities beyond isolated short term schemes.

Glasgow also attempted to stimulate its night-time activity by relaxing licensing hours in the city as part of the preparation for its role as European City of Culture in 1990. This proved problematic, however, and many of the perceived benefits, such as a reduction in alcohol-related crime, failed to materialise, and, indeed, worsened. Since then Glasgow has been forced to develop its twenty-four hour city along different lines and has abandoned its licensing initiative. More recently, Glasgow has introduced a byelaw banning the public consumption of alcohol.

Retail Initiatives

Manchester, Leeds, Nottingham, Glasgow and Sheffield have all made attempts to bridge the gap between offices closing between 5 pm and 6 pm and the start of entertainment activities between 8 pm and 9 pm. Initiatives to bridge this gap have generally been retail-based. Although Manchester achieved a participation rate in its pilot scheme of 70% of city centre retailers, Leeds and Nottingham have experienced difficulty in attracting and maintaining retailer participation. Most city centres have implemented extended hour retail schemes in conjunction with a package of measures to attract and retain people in the city. A variety of initiatives has been combined with retail, such as live music, street entertainment, theatre, cinema and restaurant promotions.

Restaurant and Café Promotion Initiatives

This is the most common initiative. The stimulation of café and restaurant activity has resulted in a series of combined schemes – with retailers or entertainment centres, such as theatres or cinemas – and usually based around the provision of discount offers. The Nottingham City Centre Nights initiative, for example, offered theatre and cinema ticket reductions. Nottingham has also gone further by developing its Guidance for the Licensing of Cafés in order to encourage and facilitate this type of development. Leeds, like many cities, has seen an increase in the number of continental-type cafés opening. Rather than any particular effort by the city council, this appears to be more due to a change in consumer tastes and consequent changes by individual establishments. Although the majority of the authorities repeatedly claim that this is part of a Twenty-Four Hour City approach, it is perhaps more indicative of a popular willingness to move towards a more continental-style of consumption in the city centre. (**Figure 11.1**)

Cardiff has adopted a more place-bound attempt to stimulate an increase in the number of the cafés within the city and has actively marketed a quarter of the city as a location for this type of development. The Mill Lane Café Quarter is a £600,000 public realm enhancement project. This urban design project has resulted in many physical improvements, including the widening of footways to allow tables to be placed outside. Cardiff anticipate that the initiative will not only have economic, city image and people-policing benefits, but will also have wider implications on land values and the re-use of vacant floorspace. This has already happened in many revitalised historic urban quarters – such as Temple

Figure 11.1 Nottingham has seen an increasing number of cafés opened. Combined with environmental improvements, this has helped establish a more welcoming ambience in the city centre.

Bar, Dublin and Shad Thames, London – where they have been the focus for a renewed desire for urbanity (Tiesdell *et al.*, 1996). The increasing demand for this type of activity has given cities a clear opportunity to capitalise upon this in a proactive way, either by direct facilitation, as in Cardiff, or by encouragement, as in Nottingham. There is a clear hope that these developments will lead to increased pedestrian density and therefore safer areas.

Although it requires a suitably opportune location – which many cities may not possess – the use of designated quarters as a vehicle for implementing a twenty-four hour approach has numerous advantages. The place-based approach to delivering the concept can be extended to areas of a city centre experiencing the loss of particular functions. A fall in rental and property values can serve to attract cultural producers and entertainment/leisure based functions, promote night time activity and contribute to the evening economy. The concentration of twenty-four hour initiatives into designated quarters has the effect of creating a zone or corridor of activity. This approach will provide a strong destination-pull and can enable resources to be concentrated on a defined area. There should also be a multiplier effect on the local economy and spin-off benefits for the other functions within the area, such as, more passing trade and overall marketing of the quarter as a destination. It is essential, however, for the approach to prove successful that a critical mass of activities and facilities is provided together with the support and provision necessary – such as public transport – to exploit these attractions.

Lighting and CCTV Schemes

Lighting and CCTV schemes are a relatively cost effective way of tackling fear of crime in the city centre at night. Lighting schemes have traditionally involved improving existing street lighting and broadening its coverage. The Twenty-Four Hour City approach, however, has developed some interesting and innovative ways of improving street lighting. Glasgow has developed a scheme with city centre retailers and Scottish Power to reduce electricity tariffs at night allowing shop window displays to remain fully lit at night, thereby bringing a degree of street illumination as well as improving store security. The importance of lighting goes beyond that of public safety, it can significantly improve the quality of CCTV picture and aid in the detection of crime. The majority of the Twenty-Four Hour City programmes involve CCTV. Leeds, for example, initiated a City Watch scheme, which is run solely by the city council.

Other Initiatives

Other initiatives within Twenty-Four Hour City strategies include the promotion of street entertainment, festivals, tourist initiatives and the promotion of the formal entertainment sector, such as theatres, art galleries, libraries and cinemas. These initiatives are aimed at creating demand for these functions by promoting links such as reduced-price restaurant meals, subsidising entry fees, or extending the opening hours, or, in the case of festivals, by actually staging the event. This multi-element approach is a good way of encouraging a more diverse range of people, in terms of age, ethnicity and gender, to come back into the city centre.

The most successful use of festivals, formal and informal entertainment, and street animation is not found in a city that is actively pursuing a formal Twenty-Four Hour City scheme. Edinburgh, while pursuing other initiatives such as its status as a European capital city, is not initiating a Twenty-Four Hour City strategy because it believes it already has a twenty-four hour city, due primarily to its large city centre residential population. Building on the success of the Edinburgh Festival, the council has attempted to further capitalise on its city image and existing tourist consumption by opening the Edinburgh Festival Theatre. Although the development of a cultural production sector of the economy is cited in the literature (Civic Trust, 1993; O'Connor, 1993) as a way of developing night-time transaction and activity, only Sheffield has actively pursued it as an initiative. The council has designated the Leadmill Quarter of the city centre for the development of the cultural industries and is actively promoting its development by designating land for cultural development.

As discussed earlier, the various Twenty-Four Hour City initiatives are aimed at local economic regeneration; enhanced city image; and improving safety. In addition they represent attempts to increase the demographic diversity of users in the city centre at night. The majority of the initiatives designed to positively attract people to the city are, however, still aimed at the youth market. Although, the initiatives are also placing strong faith in a more continental and responsible approach to alcohol consumption in city centres at night, this is a major weakness in many Twenty-Four Hour City approaches. To attract a greater range of the population to return to the city centre at night will require further consideration of this issue. Cities have yet to fully achieve the aim of encouraging a greater diversity of people, in terms of gender, ethnicity and age to use the city centre in the evening. Although the creation of a safer city centre will encourage a greater diversity of people to use the city centre, the products consumed by these groups are generally not affected by changes in the restrictions on alcohol-based entertainment.

The Twenty-Four Hour City approach is inevitably subject to fluctuations in the national economy as well as place-specific factors, both of which directly affect the demand for the Twenty-Four Hour City. The focus of all Twenty-Four Hour City initiatives, therefore, is the creation of an environment conducive to the consumption of the products of the evening economy. It is essential that the demand created is effective and actual in order to make the evening economy viable. What is particularly significant for the concept is the *additional* demand created. Arguably, however, cities are attempting to enable and facilitate the supply of Twenty-Four Hour City services on an implicit faith that a demand for such products exists. The assumption is that demand will follow supply, but if sufficient demand fails to materialise, the promotion of the evening economy will inevitably fail. Such an approach will amount to little more than a series of physical improvements that make a city centre more attractive to visit at night for a short period only. To prove successful in the longer term, the Twenty-Four Hour City approach must be an economically viable concept through all the appropriate time periods.

The residential repopulation of city centres through housing initiatives was cited by the majority of the authorities involved as one way of creating demand and implementing a successful Twenty-Four Hour City in the longer term. Although this leaves the concept open to criticisms of élitism and of the promotion of gentrification, residential development may also lead to greater effective demand for the products of the Twenty-Four Hour City. Although, given the high property values commanded by city centre locations, this might be an economic inevitability. Nevertheless, if schemes fail to open up the city centre to all of the city's citizens, it is a valid criticism.

DEVELOPING A SAFER TWENTY-FOUR HOUR CITY

The need for cities to approach Twenty-Four Hour City strategies in an holistic manner is essential. The present trading patterns for retail in the day and entertainment in the evening are well-established, and to alter them requires a major change in the culture of society. Thus, if the evening economy is to succeed it must offer something different, over and above that offered by present trading patterns. This will require innovative approaches to attract consumers and to sustain a viable evening economy, but the wider aims of the concept will not be met unless there is sufficient demand for the products of the evening economy. To create a *Twenty-Four Hour City* centre that is economically viable, safer and accessible for all the demographic groups, it is necessary to encourage both the supply of activities and enable access to them.

The fear of night-time crime in city centres is high and the image of city centres is often as places dominated by the young and by males, whose behaviour is threatening to other sections of the population. A key task for a Twenty-Four Hour City is, therefore, to reverse the trend of city centre domination by the young and reduce the perceived level of city centre fear of crime, thereby enabling it to become more accessible to all sections of the population. Greater accessibility and participation will provide the additional demand to make the Twenty-Four Hour City economically viable and safer through increased peopling of the city centre.

The many factors in the revitalisation of city centres cannot be considered as separate from the overall management of cities. The Civic Trust (1993), for example, drew attention to the fact that, like all strands of the Twenty-Four Hour City, policies and initiatives cannot operate in isolation. Particularly in the case of cultural initiatives, they should operate in conjunction with urban design strategies to provide new public spaces, and traffic calming and public transport initiatives to provide safe, accessible environments in which the activities comprising the evening economy could take place. Cities adopting the Twenty-Four Hour City approach are also attempting to generate a safer city through natural surveillance or people-policing combined with other methods of more formal surveillance, such as CCTV and lighting schemes. The creation of a safer city centre at night through the implementation of such strategies should result in the Twenty-Four Hour City and its night-time economy becoming an increasingly important part of that city's life and culture.

12

Safer City Centres in the USA

In the USA, central city decline and the decentralisation of population have been more extreme than has yet been the experience in Britain or in Europe more generally. In most developed countries the new building and construction methods and material together with the introduction of new building codes from the early 1900s, encouraged the rebuilding of city centres. In Europe, this rebuilding was largely done in situ. In American, however, it was often easier for the central business district (CBD) to migrate to a new location. Thus, in American cities without the symbolic anchor of a major plaza or cathedral square, the office core could often migrate considerable distances. American cities have therefore less allegiance and commitment to their downtown areas. By the 1970s and 1980s, the office and retail core had often migrated out of the downtown altogether. In many cities this dispersal has given rise to what Garreau (1991) has called 'Edge Cities' and what Fishman (1987, p.17) sees as 'perimeter cities' that are 'functionally independent of their urban core'. A consequence of dispersal is that most people have no need to use the traditional downtown.

Artificially high CBD land values and transport developments resulted in an erosion of the complexity and vitality of city centres, shaping them into islands of activity alive only from 9 am to 5 pm. In the 1960s and early 1970s, it was a grim joke that even muggers went in threes in the city centre streets after dark. Their retail function has suffered significantly with competition from shopping malls. Thus, between 1977 and 1984, while retail sales in America rose by 120%, retail sales in city centres fell by 50% and in many centres less than 20% of retail sales remained in the city centre (NTHP, 1988). Among the major cities, arguably only New York, Chicago, Denver, San Francisco, Portland and Seattle still have significant retailing in the city centre. Indeed, many American downtowns now no longer see retailing as their prime function, and therefore concentrate on office development and tourism, often through grand redevelopment projects. The best example of this is the Renaissance Centre in Detroit. Apart from some offices, Detroit's downtown has virtually died. With its hotel, office and lunchtime shopping facilities, the Renaissance Centre is a fortified citadel in a sea of almost total dereliction and vacancy, an area which appears unsafe and ridden with signals of crime and vandalism. Some of the vacant buildings have been cleared and, in lieu of redevelopment, trees and grass have been planted as a desperate means of beautification.

THE SPIRAL OF DECLINE

As noted in a recent article in *The Guardian* (Katz, 1996, p.3):

> The self-perpetuating spiral of white flight from America's cities is driven by a depressing logic: a city comes on hard times, so the whites who can, leave; the local economy slumps further, so businesses close or relocate, increasing unemployment; the city's tax revenue plummets, so it can no longer pay for vital services such as policing; with a shrunken tax base it is forced to put taxes up deterring new businesses from opening; with few jobs and poor law enforcement, crime soars, frightening more people away; and so it goes on.

Observing the US experience, URBED *et al.* (1994, p.36–37) argue that there are a number of steps in the process of decline. The first order effects come as peripheral centres draw off the more mobile and affluent customers, leaving the old centre dependent on a more local and poorer market. For the shops and services that remain, the most obvious response is to go 'downmarket'. Second order effects occur as the loss of trade causes stores progressively to close down and/or move. As this reduces activity and attraction, the effect can be to deter more customers from using the centre, thus discouraging the private investment needed for the centre to find a new role. A chilling example of this decline is Broadway – the main street in Gary, Indiana:

> Once Gary's bustling commercial thoroughfare, it is now a corridor of boarded up shops and derelict façades. The few storefronts still open for business are mostly charity shops or bail bondsmen. Opposite the crumbling Palace theatre, what passes for urban renewal in Gary is in progress; someone is painting a series of brightly coloured doors on to the front of a boarded up building. The old Sears department store has been converted into a welfare office. The Sheraton, last of the city's hotels, was left to the rats long ago. (Katz, 1996).

Many cities during the clearance and redevelopment programmes of the 1950s and early 1960s – the era of the 'Federal Bulldozer' – unintentionally accelerated the decline of the downtown. In many cases, revitalisation efforts failed to underpin city centre revitalisation. Buildings are neglected, fall vacant and derelict, and they eventually are cleared and not replaced, creating the so-called 'doughnut' city with a hole in the middle, as is most evident in central Detroit.

RETAIL AND CRIME

The basic requirements for successful retailing are maximum visibility, accessibility and security. Lack of security in city centres means loss of trade, which may set in motion a downward spiral. Experience in the US shows that the fear of crime impedes the ability of retailers to serve both existing and potential customers. As a city centre declines, office workers are less inclined to shop during lunch hours (Hassington, 1985). As the mobile, higher-income shoppers desert the declining city centre, the area becomes dominated by lower-income and ethnic-minority shoppers. The Citizens Commission of New York (1985) survey concluded that higher-income shoppers prefer to use retail areas used by people like themselves and they feel safer in streets with similar types of people as themselves. The finding tends to support a social segregationist rather than a social integrationist perspective. The CCC study also argued that:

> an economically vibrant downtown concentrates in a geographically compact area a wide spectrum of business, social and leisure activities. When a large number of people can walk quickly and safely from one activity to another, a downtown assumes its two unique characteristics: the capacity for visitors to engage in multi-purpose visits, known as the multiplier effect, and a high level of interpersonal communication. (CCC, 1985, p.12).

The study (CCC, 1985, p.13) showed that fear of crime damages a city centre's economy by inhibiting the behaviour of its users in the following ways:

- depressing the multiplier effect by reducing the level of pedestrian activity and the distances people are willing to walk on the streets;
- encouraging insulated activity in which self-contained complexes and in-door walkways are preferred to outside sidewalks;
- decreasing the level of face-to-face communication between users;
- promoting the desertion of the area after 5 pm;
- increasing automobile use and demand for close-by parking.

The fear of crime robs a city centre of those elements that make it a unique retail location as people either avoid coming to it altogether, especially when there are a growing number of out-of-centre shopping malls, or limit their activities significantly. Either way, retail and leisure outlets become the first losers. The New York study showed that in the USA people are most afraid of being mugged, raped or assaulted, and of having their cars broken into or stolen. Sixty-three per cent of respondents said they rarely used the city centre or walked in the streets because of fear.

Design can only create the preconditions for a safer environment: it is a poor substitute for changing the conduct of the offending individuals. But in so far as some planning policies can affect either actual crime or perceptions of safety, planners have a responsibility to explore what contribution they can make.

DESIGN AND DEVELOPMENT SOLUTIONS

There are three main strategies for dealing with the kind of fears noted in the earlier chapters of this book. The first, tackling the root causes of crime, is outside the remit of this book. The strategies that do concern planners directly may be divided into two distinct approaches: first, the provision of single-purpose urban places, segregated and protected with guards. This is the 'fortress approach', which has significantly contributed to the deterioration of the city centres and their streets in the USA and is now coming to Europe. The second approach – a Safer City approach – is to improve the city centre and its streets to make them safer. This will have a multiplier effect on all the activities of the city centre and not just benefit the retailing sector (Davies, 1982; PPS, 1984; ULI, 1983). The rest of this chapter will look at measures for creating safer streets for retail and other activities using examples from the US.

Cyril Paumier's Designing the Successful Downtown (Paumier, 1988; 1982) notes seven basic strategies essential for 'reshaping the space-use composition and economic vitality' of city centres:

- promote diversity of use;
- emphasise compactness;
- foster intensity;
- provide for accessibility;
- build a positive identity;
- ensure balance; and
- create functional linkages.

These seven principles are instrumental in building safer environments in city centres with safer streets.

In order to overcome fear of crime in downtowns, it is often necessary to create a dense, compact, multifunctional core area.

> A downtown can be designed and developed to make visitors feel that it – or a significant portion of it – is attractive and the type of place that 'respectable people' like themselves tend to frequent . . . a core downtown area that is compact, densely developed and multifunctional will concentrate people, giving them more activities . . . The activities offered in this core area will determine what 'type' of people will be strolling its sidewalks; locating offices and housing for middle- and upper-income residents in or near the core area can assure a high percentage of 'respectable', law-abiding pedestrians. Such an attractive redeveloped core area would also be large enough to affect the downtown's overall image. (Milder, 1987, p.18).

Housing and Mixed-Use Development

Some American city centres came back to life during the 1980s with the growth of gentrified areas adjoining city centres. The proliferation of restaurants and other features of 'yuppie' lifestyles allowed them to function after dark, filling the streets with pedestrians. As market forces are unlikely to produce similar results in the UK, there is a need to consider what central and local government can do to make city centres safer. Arguably the most important policy change to be considered, is how to bring housing back into city centres and make them 24-hour zones again. Experience in the USA has demonstrated the value of this policy and it is also supported by experience in most of continental Europe, where housing was never pushed out of the centre to the extent that it was in the UK.

It is often argued that pedestrians in urban streets feel safer in the presence or visual range of others (Valentine, 1990; Jacobs, 1961). City-centre residential areas are an effective way of creating well-used streets and providing natural surveillance. Even during the 1960s and early 1970s when many city centres in the USA were deserted at night, one part of Philadelphia city centre, Rittenhouse Square, continued to be a safe and elegant oasis surrounded by a diverse neighbourhood of mixed housing, clubs, art galleries, restaurants, offices, a hotel, a music school, a few small workshops, a cinema and some up-market shops. By contrast Washington Square, which was surrounded by tall office blocks, was totally deserted after 5 pm in spite of expensive redevelopment.

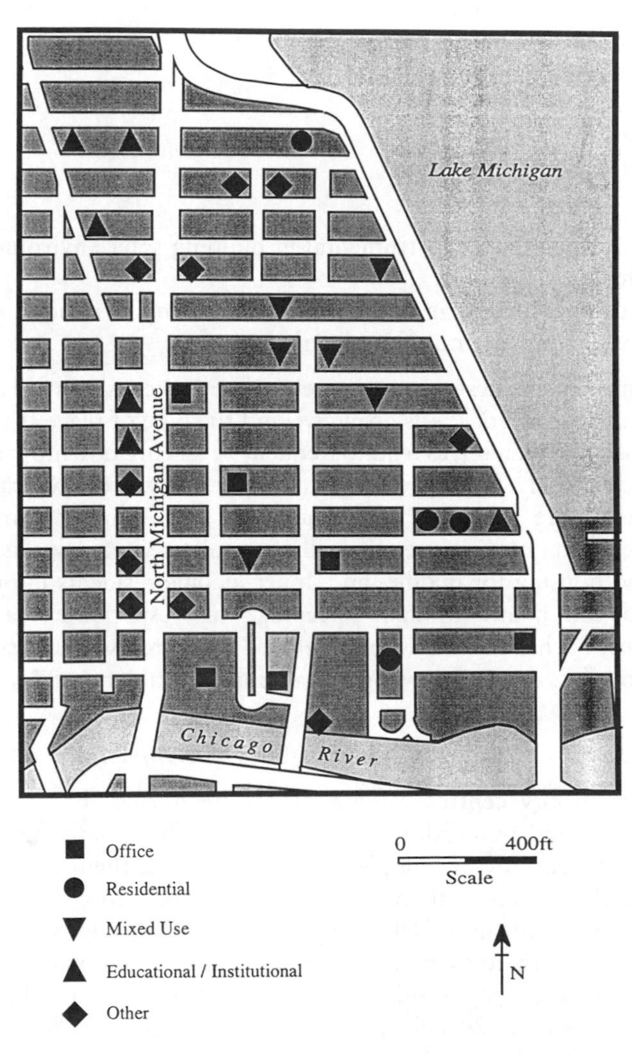

Figure 12.1 Plan of North Michigan Avenue, Chicago, showing major developments 1989–1992. The mix of activities in the area generates various pedestrian flows which ensure that the streets are busy and well-used at all times.

In the 1980s, a number of US cities experienced a dramatic change in the fortunes of their urban streets, with significant increases in the number of city-centre residents. The pattern started after 1973 with gentrification and gained momentum with the development boom of the 1980s. The most spectacular example is in Chicago. North Michigan Avenue, the shopping street for the residents of the area popularly known as the Gold Coast, used to be a muggers' paradise after the shops and offices closed. Recent changes in planning policy favouring mixed uses, however, brought 20,000 new residents into the area also known as the Magnificent Mile and its immediate vicinity (Mazur, 1991; Fieller & Peters, 1991) (**Figure 12.1**). The area has been turned into a twenty-four-hour zone occupied by residents, tourists in hotels, shoppers, diners, as well as office workers, and is one of the safest areas anywhere in the world. The crime rate along North Michigan Avenue is among the lowest in the city of Chicago.

The success of North Michigan Avenue is not based solely on its residential developments, it is also an excellent example of an 'activity corridor'. Along Michigan Avenue itself there are three large up-market shopping malls where shops stay open till 9 pm, and in two of the malls there are eating places and cinemas that stay open until midnight. The streets running off Michigan Avenue offer literally hundreds of restaurants and cafés that also stay open late and are well-used by pedestrians.

Another significant factor in making the area safe was the cleaning-up of Rush Street; the sleazy bars are gone, the go-go dancers and porno theatres are gone, and in their places are more restaurants, jazz clubs and discos, all of which are used by both sexes. The drunkards have moved away from Rush Street and the whole area around North Michigan Avenue is now much safer. While the more undesirable uses are no longer in Rush Street, they have not altogether disappeared but have been displaced to the fringes of The Loop. As long as there is a demand, activities like these will relocate. Planning policies can, however, make sure that at least some streets are free of such functions and safer for shoppers and other city centre users.

Dense and Compact Development

Housing and mixed-use developments provide the functional basis for a high level of activity in the city centre. The second measure to ensure that activity can be achieved is to plan compact developments that will increase the concentration of pedestrian movements. A busy and attractive core area minimises the fear of crime. It is necessary that there is at least a core area of the city that respectable people tend to frequent. The use of sidewalks, cafés, shops, offices by middle- and upper-middle income groups in large numbers creates a positive secure image that acts as a basis for expansion. Milder (1987) notes that the median length of a Manhattan shopping trip is only 1,200 feet and that 75 per cent of all pedestrian trips are under 2,000 feet. Thus, the diameter of a core area should usually be well under one-half mile (Schwartz, 1984).

There are several examples of downtown revitalisation schemes in the USA that are notable success stories. Michigan Avenue North in Chicago was first

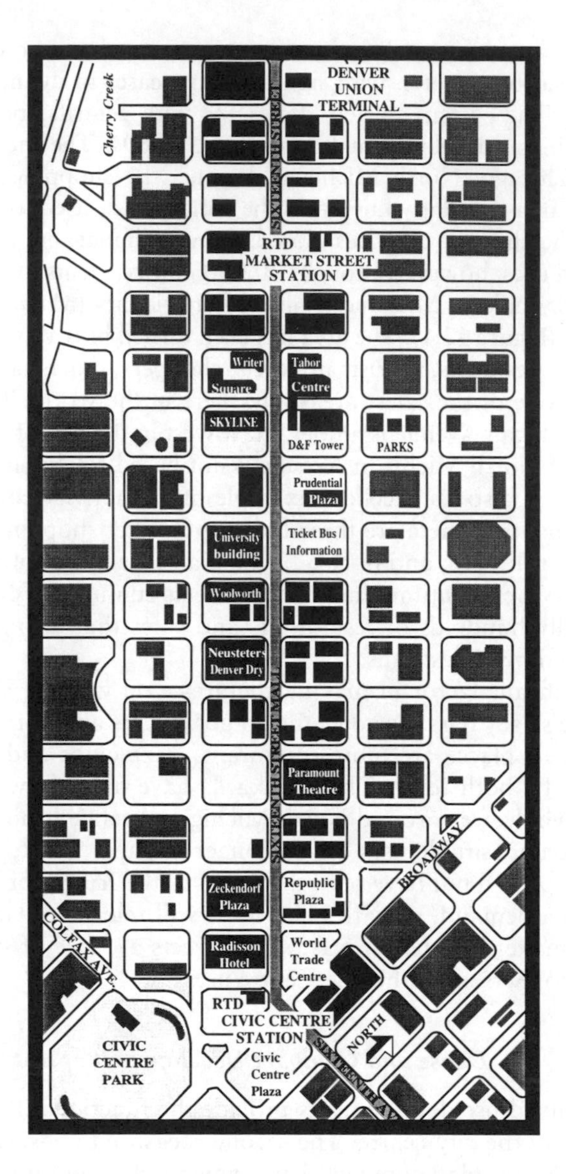

Figure 12.2 Sixteenth Street Mall, Denver. Denver's Sixteenth Street Mall is one of the best examples of an activity corridor. The Lower Downtown or LoDo area is the first four street blocks to the south east of the railway station.

established as a safe activity corridor so that developers could then be attracted to develop residential units that they would be confident of selling. The Magnificent Mile is an excellent example of an activity corridor: an avenue that is well-lit, has varied land uses, a long uninterrupted sightline, and area management policies that ensure a relatively high level of activity until after midnight. The only area which the police have to keep a particular eye on is around the West Water Tower where visibility is interrupted.

Another notable example is the 16th Street Mall in Denver, a mile-long spine running from the State Capitol and Civic Centre to Lower Downtown. As Collins *et al.* (1991, p.69) describe: 'Flanked by retail stores, office buildings and civic spaces, it represents Denver's counter attack on suburban malls and their free parking'. It is also a compact core retailing area with a very high level of daytime activity, which has created a safe and pleasant environment. Parking lots are in close proximity as they would be for a suburban mall but, unlike a suburban mall, the street is also the hub of the public transport system. The provision of a free tram service adds further to the activity and distributes it along the street. One important feature of the 16th Street Mall is that the retail enclaves developed along it are designed with open or wide-glass frontages (**Figure 12.2**). The enclosed spaces are visually an integral part of the sidewalk so that there is a constant feeling of a busy street even if some of the shoppers are on the other side of a glass panel. The 16th Street Mall is further enhanced by the revitalisation of the historic quarter – known as Lower Downtown or LoDo – adjoining it (Tiesdell *et al.*, 1996, pp159–162).

That LoDo is an integral part of the Downtown area is a major factor in its revitalisation. It benefits from – and contributes to – the Downtown revitalisation: 'These sectors: downtown retail, housing, hospitality and entertainment, must and will work in unison to come to their full potential' (Roelke, 1992, p.11). LoDo contains a mixed use of offices, retail, restaurants, housing and parking and a growing design community; all of which create a symbiotic relationship with the Downtown offices. Roelke (1992, p.3) argues that:

> LoDo's distinct character, as compared to the central core of Downtown, demonstrates the advantages of character and feel of the area. Activities, street life and shopping are important site attractions that businesses offer their employees. As a marketing tool for the central core, LoDo provides a major attractant to entice new business to Downtown. It is therefore vitally important that the success of the outer sectors occurs, in order to regain office market strength for the whole of Downtown.

LoDo is also expected to benefit further from two adjacent developments – Corrs Field Sports Arena and Elitches Amusement Park – which will increase demand for offices, entertainment, retail and residential (Tiesdell *et al.*, 1996, p.162) resulting in a safer Downtown.

One mistake in the early attempts to create compact city centres with dense activities was to build limited-access structures, such as the Trump Tower on Fifth Avenue in Manhattan or the Water Tower Place on North Michigan Avenue in Chicago. These buildings turned their backs on the streets and created well-controlled pleasant interior space in competition with – but not part of – the street. Although partially justified by climate considerations, such structures keep people from using normal sidewalk routes, especially if they are connected by off-street pedestrian networks, as is the case in Minneapolis, Toronto, Montreal and Atlanta among others. They diminish a city centre's street life and prevent visitors from engaging in more than one activity. One retailer is quoted as saying that 'they rob from the street' (CCC, 1985). By contrast, two new structures with limited access along North Michigan Avenue have opened their ground floors so that they are part of the street system. They have several entrances and ground floor shops that enhance the street with their display windows.

In designing dense and compact city-centre developments to create intensity of use, the critical mass of activity should be complemented by a balanced distribution of activity generators. A balanced distribution requires evening as well as daytime activities, and weekend as well as weekday activities. An over-concentration of office development, for example, leaves streets empty not only in the evenings but also at weekends. This is why retailing in areas dominated by offices is a weekday only activity. In Chicago, this is the case in The Loop where personal safety after dark and over the weekend becomes a major concern.

Some US cities have chosen to concentrate their safety measures on a few central activity corridors. This policy has certain benefits since those using the central area are concentrated in these corridors giving a feeling of security to shoppers and other users. Such a policy, however, has the implication that other parts of the centre may be even less safe. Wherever they are implemented, activity corridors will only work if co-ordinated with the provision of public transport services, car parks and increased provision for on-street parking.

Positive Identity

Bringing back land uses like housing and sports facilities to improve city centres takes time and requires significant policy changes. Activity corridors could be designed and implemented sooner but not immediately. Some measures, however, may be implemented quickly and with very little cost. First, improved lighting of streets creates an immediate feeling of safety. Using bright white lights instead of yellow street lamps brings a significant and relatively cheap improvement. Another measure to make streets more inviting and attract greater activity is improvement of the physical appearance of buildings and keeping streets free of litter and graffiti. Research shows that people are very sensitive to signs of deterioration in urban streets (CCC, 1985; Milder, 1987).

City-centre events are often used to build a positive image and to attract residents who do not normally use the city centres. If, however, these events take place in the early stages of improvement programmes they may have a negative effect, confirming in the minds of many that the centre is in poor physical condition and unsafe.

Safe Inner-City Communities

Another development of the late 1970s and 1980s in the USA has been the emergence of inner city communities in some cities. This has increased the number of inhabitants who use city centres and public transport in preference to shopping malls accessible by car. The often noted examples are the Capital Hill District, in Seattle, and the Castro Street area, in San Francisco. These are basically examples of urban villages. The Capital Hill area is a revitalised inner city district within easy travelling distance of downtown Seattle. It is a mixed community of yuppies and people with alternative lifestyles and centres around a safe activity corridor. The Castro Street area is different in its nature in that its origins are in the needs of the gay community to create a safe/defensible area in the inner city. Castro Street has now developed into an urban village with mixed-use developments including leisure, retail, office and residential uses. Nevertheless, it could be argued that it has only an indirect impact on the safety of San Francisco's downtown.

In the US, there are also traditional upper-middleclass neighbourhoods located close to the city centres which have had a significant impact on sustaining the retail aspect of the downtown. Society Hill in Philadelphia is one of these areas. Not only has the area maintained its resident population, it has, in fact, grown to South Street and beyond. South Street now provides an alternative, safe retail and leisure corridor for the city, underpinned by the services it provides for the residents in Society Hill. To a lesser extent, Beacon Hill and Back Bay in Boston have played a similar role. It could be argued that, apart from Capital Hill, the other three areas are 'exclusionary' communities. Society Hill and Beacon Hill exclude through income and Castro District through a certain lifestyle. Nevertheless, these communities contribute significantly to the well-being of the city centres in those cities.

Fortress Downtowns

In contrast to the developments exemplified by the Magnificent Mile, in Chicago, and the Sixteenth Street Mall and LoDo, in Denver, in the 1980s Los Angeles undertook the revitalisation of its downtown by creating what Mike Davis in his book (1990) terms Fortress LA. Davis (1990, p.229) argues that, viewed from the standpoint of its interactions with other areas in the central city, 'the "fortress effect" emerges, not as an inadvertent failure of design, but as deliberate socio-spatial strategy'.

Davis' observations are given further credence by the fact that the Los Angeles' Downtown redevelopment is tangibly different from the late 1970s and early 1980s approaches in Chicago, Denver, Boston and elsewhere. In these cities, there has been a concern and an effort to preserve the past and to revitalise it. In Los Angeles, however, the approach has been more akin to the 'Federal Bulldozer' approaches of the 1960s erasing

> all association with Downtown's past and [preventing] any articulation with the non-Anglo urbanity of its future. Everywhere on the perimeter of redevelopment this strategy takes the form of a brutal architectural edge or glacis that defines the new Downtown as a citadel vis-à-vis the rest of the central city. Los Angeles is unusual amongst major urban renewal centers in preserving, however negligently, most of its circa 1900–30 Beaux Arts commercial core. At immense public cost, the corporate headquarters and financial district was shifted from the old Broadway-Spring

corridor six blocks west to the greenfield site created by destroying the Bunker Hill residential neighborhood. (Davis, 1990, p.229–230).

The security of Downtown was ensured by cutting the traditional pedestrian links to the old centre. Davis notes that this 'radical privatisation' of Downtown's public space occurred without public debate and sees racial undertones in its conception and the disregard of democratic processes in its execution. The end result of redevelopment in Los Angeles has been to segregate the rich core of the Downtown area, known as Bunker Hill, from the decaying Broadway which has now mostly been taken over by Latino immigrants. Davis (1990, p.230) argues that the Harbour Freeway and the palisades of Bunker Hill effectively create a defensive 'moat' between the rich whites of the business world on Bunker Hill and the poor immigrants surrounding it.

Although perhaps an overreaction, in creating a defensible 'fortress downtown' protecting the private investments as well as the people working there, the authorities in Los Angeles were learning from and reacting to the Watts Riots of the 1970s. Davis (1992a, p.4) substantiates his conclusions about the motives of the authorities by pointing out that the 1991 riots 'have only seemed to vindicate the foresight of Fortress Downtown's designers. While windows were being smashed throughout the old business district along Broadway and Spring streets, Bunker Hill lived up to its name.' By flicking switches on their command consoles, security staff were able to cut off all access to the expensive real estate: bullet-proof steel doors rolled down over street-level entrances, escalators instantly stopped and electronic locks sealed off pedestrian passageways. In its special report, the *Los Angeles Business Journal* noted that the riot-tested success of corporate Downtown's defences served to stimulate further demand for new and higher levels of physical security (Davis, 1992a, p.4).

It may indeed be that in Los Angeles, as Davis (1992a, p.5) asserts, that the boundary between architecture and law enforcement has been eroded and that the LAPD has become a central player in the Downtown design process. It is also possible that, as a direct result of the city's acute problems of income differences and racial tensions which periodically flare up, the police have moved from CPTED at the micro-scale to CPTED at the macro-scale: the 'fortress downtown' being an extreme example of target hardening and access control. In addition, there is the use of CCTV. It is important, however, that in downtown Los Angeles all the necessary design measures are used to make it safer for its users walking its streets.

In Los Angeles, we see the emergence of defensible downtown because of the city's special circumstances. Yet, it might still be an exaggeration to see it, as Davis (1992a, p.8) does, as merging the 'sanctions of the criminal or civil code with land use planning' to create what Michel Foucault would recognise as 'further instances of the evolution of the "disciplinary order" of the twentieth century city'. Serious problems and threats may require the extreme solutions that have worrying consequences and alarm civil libertarians. Los Angles' 'fortress downtown', colourfully depicted by Davis, is probably a singular case where extreme measurements were needed due to the extremity of the city's problems. If the necessary steps are taken in Britain to reverse the decline of city centres then British cities will not have to resort to creating similar fortress city centres.

The fortress approach is also exemplified at a smaller scale by the compact and secure urban shopping centre, the Martin Luther King Jr Center, in the Watts district of Los Angeles. For maximum shopper security, the developers have surrounded the site with an eight-foot-high wrought iron fence, installed security cameras and floodlights, and built an observatory, as well as having infra-red detectors at the six entrances (three for cars, two for service, one for pedestrians) (Buckwalter, 1987). The centre is designed for maximum pedestrian movement in a compact area. In other words, because of the extreme violence and insecurity in the area, many of the features of an enclosed suburban mall are replicated in this shopping centre. Shoppers can come protected in their vehicles, park in a secure car park and shop in safety. This is not however an alternative to safe activity corridors in city centres, as it is mono-functional. Shoppers tend to prefer safe multi-functional centres. It should be noted that this fortress shopping centre was not created for the paranoid middle classes of Los Angeles – so often reviled by Davis – but for the poor residents of Watts who are often victims on their own streets.

CITY CENTRE MANAGEMENT APPROACHES

As well as the physical and environmental aspects of city centres, experience in the USA has demonstrated the need to consider the economic and cultural or social aspects. Organisations such as the National Main Street Centre, which supports local initiatives in smaller towns, and the International Downtown Association (IDA), which brings together those involved in the revitalisation of larger places, have distilled important lessons and promoted good practice:

> the US National Main Street Center's manuals and training programmes have been devised to deal with the four aspects of promotion, design, economic restructuring and organisation. These tend to require different kinds of expertise and involve different groups of people. (URBED *et al.*, 1994, p.71).

The US Main Street programme, for example, stresses four types of action (from URBED *et al.*, 1994, p.70):

- organisation involving building consensus and co-operation among the groups that play roles in the downtown
- promotion involving marketing the downtown's unique characteristics to shoppers, investors, new businesses, tourists and others
- design involving improving the downtown's image by enhancing its physical appearance
- economic restructuring meaning strengthening the existing economic base of the downtown while diversifying it.

The IDA (1993) emphasises that practitioners and city fathers should realise the importance of a 'shared vision' and use similar approaches to management as out-of-centre shopping complexes in terms of 'co-ordinated promotion and concern for cleanliness and security which they call Centralised Retail Management'. The management initiatives should also try to emphasise how city centres differ from the more 'artificial' and anodyne world of the shopping mall: in terms of, for example, character, the range of retail facilities, services, townscape, history and places of entertainment.

One recent and significant development in the US has been the creation of Business Improvement Districts (BIDs) which can levy special taxes on local businesses for environmental and other improvements within a defined area. BIDs can only be created after a ballot of local property owners and require a minimum 75 per cent vote in favour. As is noted in Chapter 13, several British cities are interested in the concept of BIDs.

Philadelphia was one of the first cities to establish a BID. Philadelphia's downtown had lost much of its retailing and was perceived as a particularly run-down part of the city. To provide a firm foundation, especially in terms of finance, the city was one of the first cities to introduce a BID. The establishment of the BID also led to the setting up of a central management organisation. Through a 4.5 per cent surcharge on the property tax paid by the owners of the 2,000 properties within the district, the BID raised US$7 million in its first year (URBED *et al.*, 1994, p.133). Its initial priority was to tackle problems of cleanliness and security through what has been termed a 'grime and crime' initiative. Fifty-six per cent of the first two years' budget was spent on cleaning pavements and over areas within the central area, while 34 per cent went on public safety initiatives. To implement these, over one hundred extra cleaners and forty uniformed community service reps, called rangers, were employed. Britain can also learn and try to adapt from the Business Improvement Districts which are showing positive changes in Philadelphia, New York and elsewhere. The 42nd Street BID in New York may provide the most spectacular example of rehabilitation of a very unsavoury and threatening area by realising its locational advantage.

Environmental improvements can, however, be dismissed as mere 'beautification'; a city's streets can be beautifully maintained, but remain empty. If the city's underlying lack of attraction is not addressed, then environmental improvements are, as Nick Falk (in Tiesdell & Oc, 1994, p.26) graphically states, 'like putting cosmetics on a corpse'. Nevertheless, it could be argued that safer city centres have got to tackle the 'grime and crime' afflicting them as the spiral of decline starts or – in the face of competition from other centres – to maintain their market position and prevent that spiral ever starting. Arresting decline needs management intervention, like the BIDs noted above, as well as integrated planning and design approaches noted earlier in the chapter, exemplified by Chicago and Denver.

The management approach in Philadelphia has gone beyond environmental improvements. Rangers were also recruited to undertake other improvements, such as a system of increased security, through regular patrols that provide information and through personal radios to provide early reports of crime problems. The early actions prepared the way for a more intensive programme of cultural events and activities:

> including a 'make it a night' campaign that deliberately sought to create a special atmosphere, with over 800 participating outlets, street entertainment, and cheap parking and transport for late night shopping once a week. (URBED *et al.*, 1994, p.133)

CONCLUSION

As the Policy Studies Institute (PSI) points out in *Britain in 2010* (1991), Britain must learn from the experience of North America. In North America, the process of decentralisation of city centre functions and the consequent decline of city centres rendering them unsafe has gone much further, as is illustrated by cities such as Detroit and Gary. Many are concerned that some British cities may be showing increasing signs of following patterns evident in many North American cities which have technologically protected islands of security for certain sections of their community only, while the city centre's democratic public realm has largely disappeared. It would seem logical that Britain follows the examples of continental European cities and North American cities such as Chicago, Denver, Portland, Seattle and Toronto by curbing out-of-centre developments or – at least – countering them by improving the existing city centre through new residential developments, shopping and lesiure complexes and improved pedestrian environments served by improved public transport facilities. Chicago and Denver, for example, illustrate policies which have brought people and prosperity back to city centres. These policies, learning from studies such as the Citizens Crime Commission report, started by making them safer to bring shoppers, residents and visitors back. They may be criticised for creating middle/upper income safe areas, but this has not been at the expense of the city's poorer areas.

The policy of a safe inner-city was essential for Chicago, one of the few financial centres where world-scale markets are located. The cost of an exodus on the scale of that from central Detroit would have had serious consequences for the metropolitan area, because financial centres do not migrate to the edge city, they migrate to other cities or even continents. Detroit metropolitan area survives as it was the centre of an industrial city where certain functions may decentralise or even disappear, without completely devastating the metropolitan region.

13

The Coventry and Nottingham Experience

This chapter examines two cities recognised as being at the forefront of city centre management: Coventry and Nottingham (URBED, 1996, p.1). Coventry is a city of approximately 300,000 people located to the south of Birmingham in the West Midlands. Nottingham is a city of approximately 420,000 people located in the East Midlands. As a response to local contexts, conditions and opportunities, each city has adopted a different approach to town centre management. Each however demonstrates a commitment to ensuring a safer city centre.

COVENTRY

Coventry[1] was one of the British cities most devastated by wartime bombing. The mainly eighteenth and nineteenth century buildings lining the medieval street pattern of the city centre were largely destroyed. The Cathedral of St Michael's superb fourteenth-century spire survived the bombing but the roof and much of the nave were destroyed. Plans for a comprehensive redevelopment of Coventry had been prepared in the late 1930s, but the wartime devastation gave both further impetus and opportunity. The opportunity was to rebuild Coventry city centre in accordance with the ideas of Modernism. McCrystal (1996, p.1) notes that:

> Coventry badly wanted to outpace other cities in its reconstruction plans. Its post-war central schemes were advanced more decisively than in other badly bombed cities such as Manchester or Southampton. The city was hailed as 'bold', 'brave' and 'imaginative'.

The development of Coventry's new city centre was planned in the 1940s and construction continued through the 1950s and 1960s. The centre of the new development was Broadgate, the historic centre of the city, where the old medieval north-south and east-west roads crossed. The east side was planned as a Cathedral precinct and as the site of public buildings forming the Civic

1 Much of the material on Coventry is drawn from interviews with members of the city centre management team: Peter Collard, Head of City Centre Management Team; Mark Nicholls, Assistant Head of City Centre Management Team; and Barry Cox, Senior Development Technician, City Development Directorate.

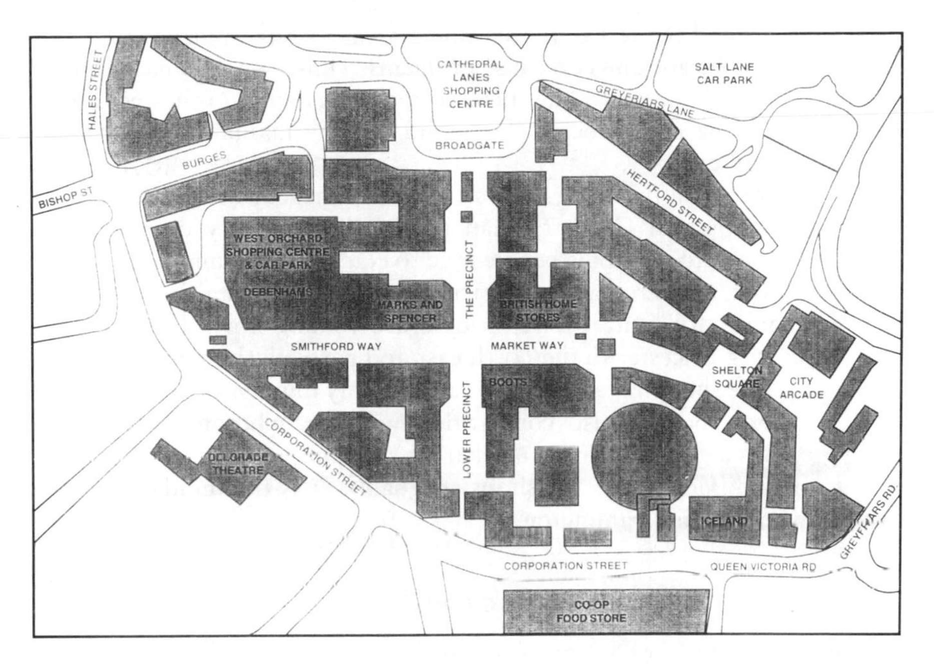

Figure 13.1 Plan of central Coventry.

Centre (**Figure 13.1**). The west side of the area was planned as the chief shopping and business centre with a new shopping precinct running east to west and with another pedestrian street at right angles to the precinct. Some of the plans for the development show this pedestrian street with traffic access based on what Percy Johnson-Marshall (1966, p.309) describes as 'the old theory of enabling shoppers to park their cars at the pavement and leave them while they did their shopping'. The City Architect, Arthur Ling, however, managed to persuade the City Council to turn the road back into a pedestrian walkway so that virtually the whole of the shopping area in the city centre was a traffic-free pedestrian precinct. During the 1950s, a new cathedral designed by Basil Spence was built around the ruins and the spire of the old cathedral. The new cathedral has acted as a magnet for the city centre and has brought in tourists and worshippers, thereby increasing the day-time uses of the city centre.

Coventry is generally a working class and blue collar city. The retail provision matches its population catchment and there are few up-market retailers. Marks & Spencer has had a presence in Coventry for a number of years and currently occupy new premises in the recently completed West Orchards enclosed shopping centre. Marks & Spencer, like most multiple retailers, generally pitches the quality of its store and the range of goods it offers based on the characteristics of the local population; the Coventry city centre branch is not a top-of-the-range store. Debenhams is one of the national retailers who are new to Coventry and located their first store in the West Orchards shopping centre. Cathedral Lanes is another enclosed shopping centre which has recently been completed and some units have not yet been let. There are also vacant units around the rest of the shopping area. As retail units are an interdependent asset, vacant units do cause problems. Thus, two or three vacancies can blight the whole street. Unlike enclosed shopping malls, it is not possible to close down a wing and relocate shops and thereby maintain a continuous frontage of retail units.

There is a perception among the city centre management team that more people are shopping in Coventry than previously. Coventry does not have major direct competitors. Some people used to come from Birmingham to shop in Coventry, but, as there have been significant environmental improvements to Birmingham's city centre, fewer people are coming now. There is some competition with Leicester to the north east and particularly with Fosse Park, an edge-of-centre shopping complex which is twenty minutes on the motorway from Coventry. There is also competition with Leamington to the south. Leamington has a generally more up-market shopping provision than Coventry and offers different types of shops. Arguably, it is the middle classes in Coventry who shop in Leamington.

Background

Although an example of the best post-war thought in terms of urban planning and architecture, as times have changed Coventry has suffered from this rebuilding. The shopping area of the city centre is a pedestrian-only precinct with its retail units turned inward onto pedestrian-only streets (**Figure 13.2**). The area is mono-functional being almost entirely retail. The upper levels were

Figure 13.2 City centre shopping area, Coventry.

designed for office uses but these are mainly vacant, while the few retail units on the upper levels have never been successful.[2] As there are very few attractions other than retail, the city centre shopping area effectively closes down after 6 pm. While the shopping centre has many similarities with covered in-town shopping malls, it is impractical for the shopping area in Coventry city centre simply to shut its gates out-of-hours and thereby exclude all visitors.

In recent years, late evening shopping has been tried but, apart from the last six Wednesdays before Christmas, it has not been a success and most shops no longer continue to open. Furthermore, as a predominantly pedestrian-only environment, there are no passing motorists to police the spaces out of shopping hours. The low density of pedestrians, therefore, makes the spaces feel unsafe. The sense of fear is exacerbated as many of the shops have external solid shutters which are pulled down at night.

As rebuilt following the destruction of the Second World War, Coventry city centre has an exceptionally low residential population. The residential population was lost in the post-war reconstruction as a result of the ideas of functional zoning being pursued. Although the University of Coventry is within the city centre, most of its fifteen to twenty thousand students live in other parts of the city. Furthermore, an exclusively student-culture in the city centre may not be widely desirable. Although within the city centre there is a theatre and an opera house, there is a dearth of restaurants and other facilities that might attract people to the city centre in the evening. People, therefore, tend to stay in the suburbs or go out of town in the evenings. As it does not really have an evening economy, Coventry city centre's principal problem is, therefore, one of day-time safety.

CITY CENTRE MANAGEMENT

Given the nature and design of the shopping area in Coventry city centre, it is logical that the centre can be managed using many – but not all – of the techniques developed in out-of-centre shopping malls. In 1986 one of the council's officers had the vision and energy to begin the process of putting a dedicated city centre management regime in place. Although many of the functions of city centre management were being performed by a number of different departments within the local authority, it was felt that there would be benefits from the co-ordination and integration of the different functions under a single regime that could focus more precisely on the city centre. The city council owns 60% of the property of the shopping area and 100% of the land, and therefore has a very direct financial interest in its vitality and viability. As part of the 'knock-on effect' of compulsory competitive tendering and the contracting out of council services, the city council has to manage its assets more explicitly and proactively. The city council is also using its property interests and land ownership to pursue partnerships with private sector developers in order to carry out physical improvements to the shopping area.

2 At the time of writing (1996), the idea of introducing living over the shop schemes was being discussed.

Two of the major actions in Coventry have been the introduction of an alcohol ban byelaw and the introduction of an extensive CCTV system in the shopping area and the city centre car parks.

The Alcohol Byelaw

Following serious public order problems with late night revellers in Coventry city centre over the Christmas and New Year's Eve period of 1984, the Coventry City Centre Alcohol Related Crime Project was established. Led by the chamber of commerce, membership of the Steering Committee was broad, including representatives from retailers; the licensed trade; brewers; police; licensing justices; city council; the local polytechnic; the Alcohol Advisory Service; and the Home Office.

Ramsey (1990, p.1) regards central Coventry in the late 1980s as 'a testing ground for a wide range of measures intended to ensure sensible patterns of drinking'. The wide-ranging series of initiatives included various types of educational material produced and distributed widely, and efforts made to improve late-night transport to facilitate the dispersal of drinkers. The police also stepped up visits to licensed premises and generally made it clear that they were keeping an eye on any alcohol-related problems. Against this background, the Committee developed its most dramatic proposal, for the city centre to become an 'alcohol-free' zone. At the time, Coventry had a major problem with 'lager louts' and 'winos'. A survey revealed that 12% of visitors to the city had been affected by drink-related behaviour, a consequence of which was to deter them from using certain parts of the city centre (URBED *et al.*, 1994, p.117). There were also aspects of a *zeitgeist*. As Ramsey (1990, p.2) notes 1988 was '"the year of the lagerlout", marked by major concern in the media and elsewhere over the disorderly antics of hard-drinking young men, soused in lager'. The city centre management team readily admit that at the time Coventry was a desperate place.

Thus, in 1988 Coventry City Council became the first local authority in Britain to introduce a byelaw prohibiting the public consumption of alcohol in its city centre streets. The byelaw was introduced after consultation with local property owners and landlords and publicans, from whom there was universal support; the problem of alcohol-related disorder being 'widely perceived by local people as needing serious attention' (Ramsey, 1990, p.25). The Home Office opted for a two-year experiment to test Coventry's byelaw in practice. Although six other authorities were involved in the experiment, Coventry was the flagship. Coventry's lead has since been followed by a number of other authorities, most recently and notably Glasgow. Furthermore, in September 1996, the Home Office (1996i) also announced proposals to give the police powers to confiscate alcohol from under-age drinkers in public places.

To examine the impact of the byelaw, two surveys (Ramsey, 1990) were carried out in central Coventry, one shortly before the byelaw took effect in November 1988 and another a year later. The 'before' survey found that although many people feared that they might be victimised on their visits to central Coventry, the actual experiences of victimisation were comparatively

Figure 13.3 Alcohol ban byelaw sign, Coventry. Coventry has a very distinct inner ring-road which forms the boundary for its alcohol ban byelaw.

rare – about 2% of the respondents claiming victimisation; significantly fewer than the 12% of the earlier survey. The survey did reveal that

> 59 per cent of the interviewees said that they sometimes avoided certain sorts of people – 'drunks, winos and tramps' being the most commonly cited category. Substantially fewer reported avoiding 'groups of people' or 'young men', or anyone else . . . [Thus, although] . . . only quite small groups of drinkers gathered in the city centre – and then just intermittently – they were evidently rather conspicuous, and a focus for fear. (Ramsey, 1990, p.4).

The byelaw covers the area within the ring-road and it is the police who are responsible for enforcing the byelaw (**Figure 13.3**). Those seen drinking are asked to desist, discard the drink or to move on. Since the byelaw was introduced, there have only been seven prosecutions for drinking. This is, in part, because the actual offence is not drinking, but continuing to drink having been asked to stop by a police officer. Ramsey's research found that several police officers were aware that enforcement of the byelaw depended on the public having an exaggerated perception of police powers, such as powers of arrest, under the byelaw (Ramsey, 1990, p.21).

The Impact of the Byelaw

Police statistics for recorded crimes in the city centre showed that the key categories of assaults, robberies/thefts from the person and criminal damage, appear to have been unaffected by the introduction of the byelaw (Ramsey, 1990, p.5). The surveys did however reveal a reduction in incivilities following the introduction of the byelaw. The proportion of interviewees who saw public drinkers as a common problem was reduced from over half in the 'before' survey to less than a quarter in the second survey.

Both surveys recorded high levels of popular support for the byelaw. Ramsey's research also shows that while there had been a high level of popular support for the byelaw (86%) prior to its implementation, the 16–30 age group were less wholehearted in their support (76%). After implementation, the general support for the byelaw increased (93%) but so, to a greater extent, did the support from within the 16–30 group (86%). Ramsey's research (1990, p.22) noted some displacement with 'small groups of drinkers congregating in . . . a particular park outside the ring-road, as various officers conceded'. Discounting these other social costs, there has been the positive achievement of making the city centre perceptibly safer. The city centre management team however stated that currently the winos seem to have disappeared from Coventry.

There are two important caveats to the experience in Coventry (Ramsey, 1990, p.118–19). First, the introduction of all-day drinking in pubs, a few months prior to the start of the byelaw, probably also contributed to the byelaw's success. Secondly, the 'alcohol-free zone' was limited in size and clearly demarcated by the ring-road which together with warning notices, meant that anyone aware of the byelaw knew where it did and did not apply. In accordance with the situational crime prevention technique of 'rule setting' this removes ambiguity. It also means, however, that its success may not be transferable to less well-defined or more extensive areas.

More recently, by inhibiting the development of a 'street-café society' in the city centre, the success of the alcohol ban, in combination with other actions, has created a different kind of problem. When the byelaw was introduced there was no vision or expectation of the possibility of a 'café society' in Coventry despite it being 'the British city which also pioneered the pedestrian precinct' (Arnot, 1995, p.12). The city council is in the process of applying to the Home Office to have the byelaw amended to permit the public consumption of alcohol in certain limited and controlled circumstances: for example, in designated areas adjacent to licensed premises with waitress or waiter service. Cities that have introduced the alcohol ban byelaw after Coventry generally included this provision.

In addition to byelaws, as the city council owns the land in the shopping area, it can impose controls and restrictions on the use of the land in a manner analogous to enclosed out-of-town shopping centres. Effectively, it has greater powers in this respect than most local authorities. There are already controls, for example, on certain activities such as street entertainers, charities, and political groups. The city council has the power to control access and could, for example, ban known shoplifters from entering the shopping area. There is a famous local case of a troublesome family being banned from a housing estate in Coventry. The police have advised the council that legally it could do the same in the city centre, but for largely political reasons the council has not deemed it expedient to follow this option. Another measure within the central shopping area is a Retail Radio Link between about forty shops, to which the police are also connected. This enables the rapid relay of information about, for example, gangs of shop-lifters operating.

CCTV Cameras

The second significant action in Coventry city centre has been the installation of a closed circuit television system. In 1987, the first ten cameras were installed; three of these were in car parks, the rest were in the shopping area. By 1996, there were over one hundred cameras; two-thirds of which are in the car parks. Some cameras can survey both the shopping area and the car parks. There has been no attempt to conceal the cameras; the city council wants people to know that the cameras are there. The installation of the cameras was originally presented as a management tool in order to guarantee the safety of people using the car parks and shopping area. Many of the car parks have signs advertising the presence of the cameras. This also enhances their deterrent effect. The city centre management team argue that people are reassured by knowing of their existence, but otherwise do not regard the cameras as intrusive.

At present members of the city centre management team operate the monitors, with typical shifts of ninety minutes. The shifts are limited based on the recommendation that two hours is the maximum time that people should use computer screens without a break.[3] Nevertheless, there are just two operators at any one time using about twenty monitors to control over one hundred cameras all with pan, tilt and zone. The efficiency of the system is crucially dependent on the skill of the operators and their ability to process a potentially enormous amount of visual information. The operators develop knowledge of potential trouble spots and times and can thereby focus their attention. In case any incident is missed – and although it does not help to deal with the incident when it happens – a time-lapse recording is made for all the cameras. There are also blind spots where the cameras cannot see. There is a conflict between the demands of visibility and surveillance and that of aesthetics. The city centre management team would like to cut down trees and trim hedges and bushes back, while the planners would like to keep them. The need is therefore to reconcile both visual amenity and safety.

3 Evidence from the West Midlands Police (1994, from Beck & Willis, 1995, p.188) suggests that an operator's effective span of attention is less than twenty minutes. After this, a rest of equal length is required.

The CCTV system has a watching brief; dealing with any criminal or potentially criminal offence or incident is seen as a job for the police. The CCTV system monitors a range of different kinds of 'incident' from vandalism and buskers to lost children. It also monitors and forms a control point for parades and processions. The city centre management team assert that there is 'a small number of known thieves' who are responsible for a large proportion of incidents, and keep a specific vigilance on this group. The team are, however, careful to point out that in operating the CCTV system they have to be careful to avoid infringing people's civil liberties. There are very strict guidelines imposed by the council on the use of the cameras and on the use of any tapes produced.

Currently the system is only monitored until 10.30 pm and there is only an indirect link with the police. The operators call in the police when they think there is an incident which requires police attention. A direct link with the police will soon be established and, using money from a successful bid for Home Office CCTV Challenge, the police will take over the night-time operation of the system. The police however will not have the manpower to monitor them continuously.

In one of the car parks, as a short term measure, dummy cameras were installed. The cameras had the desired effect and reduced the crime rate. When an incident was reported and the offender demanded to see evidence of his offence, the bluff was called. The city centre management team insist that dummy cameras will not be used again: people need to trust that someone is

Figure 13.4 Improvements to car parks in Coventry city centre. Improvements have included the installation of CCTV cameras, increased visibility and lighting (for example, full glass doors into lobbies; perimeter walls lowered to give views into the car park from the street; hedges and bushes cut down), and other environmental improvements to create a cleaner, brighter appearance (for example, painting surfaces and ceilings white).

watching and dummy cameras could undermine both public confidence and the credibility of the system.

Safer Car Parking

Car crime and crime in the various city centre car parks had been a major problem in Coventry, contributing significantly to the negative perceptions of the city centre. There are 6500 city centre car parking spaces in Coventry city centre in six multi-storey and nineteen surface level car parks, all of which are council-owned. There has been a series of improvements (**Figure 13.4**). The intention has been to reduce the opportunities for crime. As a result, car crime has been reduced by 60% over ten years and usage of the car parks has gone up by 30% in the same period. A senior local police officer has said that it is now safer to park in the city centre car parks than outside your own home. There have been some conflicts with the local planners who tend to want to ameliorate the visual impact of car parks by hiding them. Hiding the car parks, however, makes those using them and their cars more vulnerable.

Two of the 26 car parks are manned in the city centre. Attendants also walk around and may cover several car parks. The intention in Coventry has been to make all of the car parks safe for everybody. Thus, there are no 'women-only' car parks. The police would like to see some of the car parks closed in the evening to concentrate usage and make the remaining car parks safer. The city centre management team, however, argue that people will want to use the most convenient car parks and do not like to walk. Closing some car parks may mean they have to walk through less busy – and therefore – potentially more dangerous areas to get where they want to go. If most of the car parks are open, people can trade-off convenience against security and safety. Signs advertising theatres and sports centres also suggest where to park.

Environmental Improvements

There are certain design problems within the shopping area. A rolling programme of improvements is currently being undertaken with important safety implications. The improvements involve increasing visibility and removing many of the 'nooks and crannies' which people were concerned about and where potential attackers could hide. The Lower Precinct with its Round Café precinct and perforated balustrades is the last section to be rehabilitated. The Round Café itself is the source of some local nostalgia (**Figure 13.5**). The project is a partnership between the city council and private sector developers. There has been some conflict with English Heritage who regard the precinct as a fine example of *Festival of Britain* architecture and, accordingly, have been reluctant to see major changes. English Heritage initially threatened the possibility of 'overnight' listing, unless the city council and the developers complied with its recommendations, but this gave way to a more constructive dialogue.

Figure 13.5 The Lower Precinct, Coventry city centre. This will be the final phase of the current programme of improvements to Coventry city centre. The Round Café is the source of some local nostalgia. The precinct is considered to be a good example of Festival of Britain architecture.

THE FUTURE

Coventry will be changing the structure of its city centre management approach and has completed negotiations with private sector partners to set up its innovative city-centre management company. Although similar schemes operate already in the US, the project goes further than other city-centre initiatives in the UK. The House of Commons Environment Committee (1994) had previously discussed the American idea of BIDs (see Chapter 12). Rather than a compulsory local tax, the Committee had suggested funding them by a 'City-Challenge-type scheme' through the Single Regeneration Budget or by extending tax relief on contributions made by private companies to approved town-centre initiatives. The government is reluctant to be seen to be imposing new taxes and any contribution from the private sector would need to be voluntary. A voluntary levy, however, raises the problem of 'free-riders' who gain the benefits of the improvements without contributing to the costs. Research has recently been jointly commissioned by the ATCM and the DoE on the idea of Town Improvement Zones (TIZs).

Coventry's Labour-run city council will be the first in Britain to hand over management of its whole city centre to a private company, the Coventry and Warwickshire Partnership. This is a joint public/private sector company with its own revenue-raising powers, with limited company status. The management company's board members will include: national directors of two property firms, a national director of Boots, a local vicar, a hotel manager, a university vice-chancellor, and the chairman of the local technical college. It also includes a city councillor and a council officer (UET, 1996, p.1). Following the appointment of a Chief Executive, the company was expected to take over management of the city centre by the autumn of 1996. In Coventry, the company will initially have 'janitorial' responsibilities looking after the city's multi-storey and surface council-owned car parks; litter removal; enhancement projects, such as paving and landscaping, and running the CCTV system. To carry out many of these tasks, the company is also expected to employ seventy-five city centre 'ambassadors'. As there will not be any transfer of assets from the city council control, the city council will still remain accountable for the performance of the company.

Nottingham's city centre manager, Jane Ellis, is sceptical of the advantages of a city centre management company. She argues that city centre managers are most effective as facilitators rather than as direct providers. This concurs with a contemporary view of local authorities not as direct providers, but as facilitators providing a framework within which commercial interests operate. She argues that, in effect, Coventry's city centre management company is a return to the older notion of direct provision; the management company resembles a single-purpose, mini-local authority (albeit with only indirect local democratic accountability).

NOTTINGHAM

Nottingham[4] has been a major city in Britain for a number of centuries. Already a large city prior to the Industrial Revolution, it expanded rapidly in terms of population during the middle years of the nineteenth century and continued to be a major light manufacturing centre into the twentieth century. The current population of the city of Nottingham is approximately 420,000; a figure which does not include the very large suburb of West Bridgford. The population of the Greater Nottingham travel-to-work area is 750,000, while the population of Nottingham city centre's retail catchment area is estimated to be about two million people. In the twenty-five years since the opening of its two city centre shopping centres, the introduction of the central core traffic schemes and, more recently, pedestrianised areas and other environmental improvements, the city centre has developed to become a major regional shopping centre. It contains an estimated 1.7 million square feet of net retail floorspace and, in terms of the number of multiple retailers represented, it is now the fourth largest retail centre in the country (**Figure 13.6**).

The city centre has a wide range of shopping. The Broad Marsh Centre, on the southern edge of the city centre shopping area, is a mass-market shopping centre. With the recent addition of an Allders department store, it has gone more up-market. The Victoria Centre on the northern edge of the shopping area is a middle-market shopping centre. Both of these shopping centres date from the late 1960s and early 1970s and although still handicapped by their design, both have recently been refurbished. The two shopping centres are

4 Much of the material on Nottingham is drawn from interviews with Jane Ellis, the current Nottingham City Centre Manager, and Martin Garratt, the former Nottingham City Centre Manager (now Town Centre Support Project Manager with Boots The Chemists).

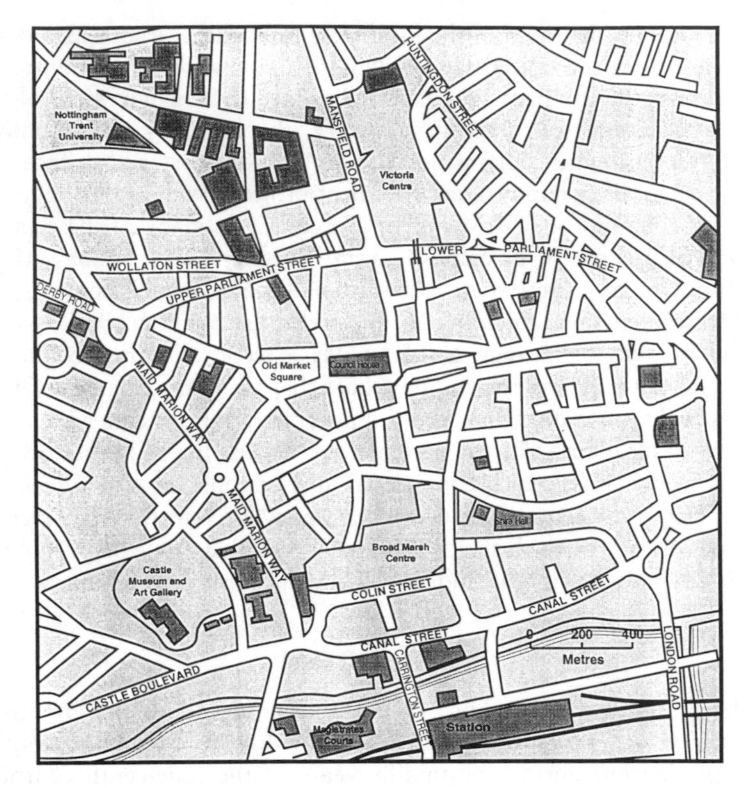

Figure 13.6 Plan of central Nottingham.

linked by the pedestrianised streets of Clumber Street and Bridlesmith Gate. In terms of pedestrian flow, Clumber Street is one of the busiest shopping streets in Europe. A new arcade, Flying Horse Walk, and the refurbished Exchange Arcade, both in the centre of the shopping area, together with Bridlesmith Gate cater for the speciality and upper end of the retail market. Bridlesmith Gate itself is rapidly becoming one of the most elegant shopping streets in the UK. The day-time retail economy is on a north-south axis (Victoria Centre to Broad Marsh) terminated at each pole by one of the shopping centres. It is also significant that Broad Marsh and Victoria Centre do not directly compete with the streets in terms of pedestrian circulation. Thus, the city centre's permeability is not significantly diminished when these are closed at night. By contrast, the evening leisure economy is on a more east-west axis (from Chapel Bar to Hockley and the Lace Market) (**Figure 13.7**).

Nottingham's trading position is however under challenge. Sheffield's Meadowhall regional shopping centre is less than an hour's drive north up the M1 motorway. The manageable drive-time makes it an attractive alternative to Nottingham's city centre. In addition, the city centres of Sheffield, Leicester and Derby have all recently benefited from the completion of major new city centre shopping developments and smaller towns, such as Loughborough, Lincoln, Mansfield and Newark, are also developing their range of retail facilities.

Figure 13.7 The Market Square, Nottingham.

Nottingham also has some weaknesses. These include its poor food shopping offer; some major store branches not being in their company's top category; poor access, congestion; perceived shortage of car parking; perceived high charges relative to free out-of-centre parking; and the poor quality pedestrian/shopping environment of the inner ring-road. In addition, the city centre in the evenings is dominated by a high proportion of young people which makes the centre less attractive to other groups. Although the city centre is served by a ring of twelve multi-storey car parks all located within or near the inner-ring road, there remains a strong public perception that it is difficult to park near the city centre. Given the abundance of free and convenient car parking at Meadowhall for example, the city council recognises the need to improve provision and equally important to change people's perceptions about the ease of parking.

CITY CENTRE MANAGEMENT

By the early 1990s, Nottingham City Council was regarded as being good at developing strategic visions through its Local Plan and various planning and planning-related policies: revitalising the city centre; extending the Broad Marsh shopping centre, encouraging mixed uses; encouraging residential uses in the city's historic Lace Market, etc.. There was, however, a realisation that there was still a need for day-to-day management. This lack was seen as a deterrent to investment. Within the City Council, there was no extra money available for this and a city centre management strategy was seen as a way of enabling extra funds to be brought in with the private sector as co-funders.

A City Centre Management Steering Group was set up, consisting of representatives from Nottingham City Council; Nottinghamshire County Council; Boots The Chemists; Marks & Spencer; the City Centre Retailers Association; Nottinghamshire Police; the Nottingham Safer Cities Project; Nottingham Development Enterprise; the Victoria Shopping Centre; the Broad Marsh Shopping Centre; and Dusco Ltd.. This group met monthly through 1990–91, with much of the discussion focusing on determining the framework under which a future city centre manager would work. Safety in the city centre was one of the Steering Group's key concerns.

The Steering Group was able to use various consultancy reports which the City Council and other bodies, such as the Nottingham Safer Cities Project, had commissioned during the late 1980s and early 1990s. These had examined such factors as: footfall counts around the city centre; competitor analysis (including the possible impact of Meadowhall Regional Shopping Centre); a strengths, weaknesses, opportunities and threats (SWOT) analysis of the city centre; a profile of inward investors, and crucially in this context, a study of people's fear of crime and their usage of the city centre.

The Nottingham Safer Cities Project

The city centre management effort was able to build on the foundations of the Nottingham Safer Cities project. The Nottingham Safer Cities Project's Action Plan of 1989/90 had three main priorities: to make the city centre safer; action in local areas to improve safety and reduce crime, and action to help the groups who suffer most from crime. In terms of making the city centre safer, the proposal was to commission a Community Safety Plan for the city centre. A Working Group was set up to develop a community safety policy for the city centre, involving the local authorities, city centre retailers, commercial and leisure interests. The first priority was to improve safety in the 'gateways' into Nottingham – the bus stations and car parks. A second priority was to tackle drink-related offending in the city centre involving offender based and pub-based proposals, and teaching young people how to enjoy alcohol without causing offences to other people (Nottingham Safer Cities Project, 1989).

The Safer Cities Project also established five principles of good design and management to improve safety within the city centre. These were:

- design for pedestrians to move about in well-lit, wide circulation routes which reflect patterns of movement
- consider the safety of people and property together rather than separately
- use opportunities for enhancing natural surveillance
- ensure good maintenance to avoid giving clues of decay and neglect to the casual criminal
- make sure the solution to one problem does not create another

(Nottingham Safer Cities Project, 1990, p.7–8).

The Nottingham Safer Cities Project also commissioned the Nottingham Crime Audit (1991) undertaken by KPMG Peat Marwick and the Safe Neighbourhood Unit between December 1989 and July 1990. The intention was to examine the nature and costs of crime within the city and to provide a basis for new approaches to improving community safety and reducing crime. Using known levels of city-centre spending and survey indicators of the number of people avoiding the city centre through fear of crime, the value of lost sales and lost profit by reason of shoppers' avoidance behaviour was estimated. Although the method of calculation employed is open to debate, a figure of £12 million in lost turnover during the six months of the study was identified. It was estimated that this represented some £840,000 in lost profit and was equivalent to the loss of 462 job opportunities. The report's overall conclusion was that crime was an 'unavoidable fact of life' in the city, creating concern and intruding into local people's lives, absorbing public and private resources, had an adverse impact on the attraction of new employment and inward investment and threatened continued recognition of the city as: 'a major retailing . . . centre in the face of competition from newer, out-of-town centres' (KPMG Peat Marwick, 1991, p.5, from Beck & Willis, 1995, p.51).

The City Centre Management Business Plan

In 1992, a city centre manager and assistant were appointed and a city centre management Business Plan formulated listing a number of objectives. The objectives were in priority order as voted on by the members of the Steering Group. Safety was the fourth objective after accessibility; promotion of the city centre; and environmental improvements. The intention was to design the programme of projects, implement them and then to monitor the impact. Work on the other objectives also had a bearing on crime and safety: for example, Car Park Watch and Going for Gold – to encourage car park operators to make improvements and apply for recognition of quality under the AA/ACPO Secured Car Park scheme – were both under the accessibility objective.

The safety objective was 'to monitor crime levels and implement a strategy to reduce crime and the fear of crime in the city centre' (Nottingham City Centre Management, 1993, p.9). Within this the Action Plan had two more specific policy intentions, first, to 'liaise with the Police, Safer Cities and the local authorities to understand crime levels and implement agreed action plans to target specific crime problems'; and secondly, to 'encourage increased and varied activity in the town centre outside normal shopping hours' (Nottingham City Centre Management, 1993, p.35).

Five specific projects were listed under this objective. First, to commission an implementation plan for the Community Safety Plan for the city centre and to target specific problems. Secondly, to improve the day-time atmosphere in the city centre encouraging open air cafés similar to those found in other European cities. Thirdly, to review the control exercised over charity collectors and other pamphleteers. Management of these groups would be improved ensuring that disruption and intimidation of shoppers is kept to a minimum. Fourthly, to establish a day-time Shop Watch scheme; and fifthly, to market positive crime trends to raise awareness of successes in reducing crime.

Community Safety Action Plan

One of the most important projects was to commission a Community Safety Action Plan. This was undertaken by the University of Leicester's Centre for the Study of Public Order (CSPO, 1993). The CSPO made forty-nine detailed recommendations for improving personal safety in the city centre. The first twelve concerned subways; the first of which stated that all pedestrian subways should be closed and replaced by crossings at ground level. Nevertheless, it recognised that this could not be done immediately and suggested a number of ways to improve them in the interim. The presence of subways has been a particular problem in Nottingham: to the west of the city centre shopping area under Maid Marian Way; to the north under Upper Parliament Street, particularly in front of the Theatre Royal; and to the south linking the railway station to the Broad Marsh shopping centre. The city council's strategic vision is to get rid of all the subways and introduce surface level crossings. Near the Broad Marsh Centre, the scheme to do this has already started. In the shorter term, however, they remain a problem. Some people argue that it is better to let them rot in order to create the political will to get rid of them sooner. The city centre management team has tried to make efforts to improve them. In front of the Theatre Royal, the theatre was persuaded to adopt the subway in return for the use of some advertising hoardings within the subway. The battered metal ceiling was removed and the whole subway repainted white to freshen it up and improve visibility. Local university students painted murals and the cleansing regime was improved.

Recommendations 13–22 concerned 'Consumption, Leisure and Entertainment', one of which stated:

> It is imperative that CCM, together with police, retailers, licensing industry and media form a working group to discuss and agree a policy to improve and diversify current city centre usage.

Recommendations 23–44 concerned 'Informal and Formal Methods of Policing', including 'a public opinion survey to determine whether there is demand for increasing the visibility of city centre police provision during the day time'. The final five recommendations concerned 'Positive Marketing of the Civic Realm'.

Creating the Right Ambience

The intention in Nottingham was to present a very positive image with the hope that good impressions would crowd out bad. The city centre management team, for example, felt that signs, as in Coventry, banning the public consumption of alcohol would give a negative perception of the city. Instead, what was required was to project a positive image. This also required good housekeeping, reducing physical incivilities, ensuring that the physical environment was kept in good repair and that graffiti and other signs of neglect were quickly attended to.

One of the city centre manager's main tasks was to improve the media's presentation and images of the city centre. The presentation in the media tended to emphasise negative perceptions of the city centre. The publication of the Community Safety Action Plan, for example, had been greeted by headlines which gave the impression that such a plan was desperately needed as the city centre had been overrun by yobs and louts (*Nottingham Evening Post*, 1992). Although meetings were held with representatives of the local media, it was felt that there was little that could be done directly to change the media's presentations of events. Instead, a positive campaign was launched to provide a good supply of positive copy about events in and aspects of the city centre to offset the bad stories. To this end, a commercial director from Bass Leisure was seconded to work with the city centre management team one day a week for six months.

Car Parks

The first area of crime to be addressed in Nottingham city centre was car crime, which made up a quarter of all city centre offences. Two-thirds of the people surveyed for the Crime Audit were worried about car crime and retailers had also requested action as car crime was seen as deterring potential customers from coming to Nottingham. One of Nottingham's particular problems is that the ownership of the car parks is split between nine different owners (the city and county council plus seven private owners). One of the city centre manager's initial actions was to convene a forum of the different owners, to embarrass them by showing them the comparative police crime figures for their car parks and to challenge them to do something about it. A Car Park Watch scheme was introduced to help combat gangs stealing cars – known to the police as 'twoc-kers' (taking without owner's consent). Previously, when the gang had been deterred in one city centre car park, they would move on to another one; now the car park operators could warn each other that a gang was operating. In addition, the car park attendants were given training in how to give descriptions of offenders that would be useful to the police.

In terms of physical improvements, the city centre manager arranged for the local Police Architectural Liaison Officer (ALO) to visit each car park and to make recommendations on improving safety. This included painting the car parks white to improve visibility and the installation of CCTV in some locations. Where CCTV cameras have been installed they link back to the car park attendant's office. Car park owners were also given an additional stimulus to improve their car parks by the introduction of the national 'secured car park' award scheme. The Trinity Square and Broad Marsh car parks both won the Gold Award. Over a nine month period, usage of the car parks increased and there was a 36% reduction in crime.

The various actions taken to improve Nottingham's city centre car parks have reduced car crime to – in the words of the current city centre manager – 'almost nil'; 'twoc-kers' in particular have gone. The most marked example is the Broad Marsh car park. In 1992 and 1993, there were 183 and 121 auto-crime offences. After installation of CCTV, this had been reduced to, in 1994 and 1995, thirty and nineteen offences. For the year up until March 1996, there had been no reported auto-crime offences in this location (Chief Superintendent Richard

Dillon, from Nottingham City Centre Management, 1996, Appendix A). It is noted, however, that car crime seems to have been displaced to out-of-centre supermarkets and hypermarkets and to the city council's park-and-ride car parks. The perception of safer car parks has also enhanced the evening economy of the city. Most of the car parks do shut overnight, although some of the principal car parks for the entertainment areas stay open late or all night. Trinity Square car park stays open all night, while Fletcher Gate car park is open until about 2.30 pm.

Retail Crime

As car crime seems to have been addressed, the current major areas of crime are retail crime and public disorder. To address retail crime, retailers were encouraged to install CCTV systems. A Shop Watch system with radio link was also established. This involves about forty stores who share information and assistance, for example, personnel may be sent from one store to a particular store which needs to deal with a problem. The participating stores are mainly clustered in the Broad Marsh Shopping Centre and Listergate area as this is where the major problems were. It has been a problem to get retailers to acknowledge that they have a common problem and would be more effective in dealing with it by working together. As Beck & Willis (1994, p.83) argues the starting point is for retailers 'to acknowledge the extent of the problem and to accept that "owning" the problem is a necessary condition for responding to it'. In response to public request, the visible police presence in the pedestrianised areas of the city has also been increased.

There are two types of shoplifting happening in Nottingham. The first is relatively small-scale and opportunistic. The police suggest that this is often to feed drug and drink problems. The second is from organised gangs of shoplifters stealing several thousand pounds of stock per day. The presence of such gangs has been a problem in Nottingham for about a year and it is thought that their presence is a result of displacement from London.[5]

The Evening Economy

Nottingham's evening economy is one of its principal differences with Coventry. Suggestions for improving the evening economy had been the final part of the CSPO report. Although the fact that Nottingham has between 20,000 and 30,000 visitors on Friday and Saturday nights could be considered a successful evening economy, there were concerns that this crowd was narrowly concentrated within the 16 to 25 age group and that their presence actually deterred

5 The Marks & Spencers store in Oxford Street had a particular problem with organised gangs shoplifting. They undertook a Retail Crime Initiative involving the installation of a CCTV system and started to build up a picture library and dossiers on known shoplifters. Marks & Spencers also began to circulate and exchange their information with other stores to build a common front against the shoplifting gangs. The scheme was so successful in London that the gangs sought opportunities elsewhere, initially in the rest of London and then further afield. This has seen a national displacement of the problem.

other age groups. After lamenting the state of the city centre, the editorial in the *Nottingham Evening Post* (1992, p.4) accompanying the publication of the Community Safety Action Plan, agreed with the proposed approach:

> We accept entirely that the majority of the 20,000 revellers to be found in the city on Friday and Saturday nights are decent, normal high-spirited youngsters who no doubt find the brainless gangs of louts just as offensive and 'naff' as their elders do. But nocturnal enjoyment of our very attractive city centre should not be the prerogative of one age group. And the removal of the 'mono culture' would eventually force decent standards of behaviour on the yobs because they would no longer have a domineering presence. (*Nottingham Evening Post*, 1992, p.4).

The city centre management team's intention was to promote positive images of the city centre's night life and to broaden the catchment. Recognising that the social culture of the country could not be changed overnight, the intention was to develop the evening economy gradually. The effort started by focusing on late night shopping on Wednesday evenings. A number of shops were persuaded to stay open late and this was combined with a number of special offers, such as reduced price meals; offers on cinema tickets (four for the price of three); live music in restaurants; cheaper car parking. The city has also tried to encourage the development of a café-culture through the development of cafés, particularly those with outdoor seating. The policy was one of explicitly 'Europeanising' the city.

To complement its Shop Watch scheme, Nottingham also has a Pub Watch and a Club Watch which enable managers to exchange information on possible problems. Most cities now have Pub Watch schemes, fewer have Club Watch schemes as cities need a critical concentration of clubs to make it worthwhile. The scheme is driven by the larger, more mainstream clubs but the smaller clubs are also keen to be involved. The police also issue regular information updates to both networks, including statements on which people have been barred from clubs or pubs by magistrates and when their sentences have been served. The various Watch schemes are examples of people and organisations recognising that their individual interests are better served by working together.

Public Order

In a Perceptions Study, undertaken in 1995 (Social Research Associates, 1995), 70% of respondents considered that the city centre had improved in recent years. The survey found that of fourteen identified problems, the two which prompted most concern were night-time safety (38.5% of all respondents) and crime (28.5%). Parking was ranked third (25%), while day-time safety was ranked seventh (11%). Among the 30% who did not regard the city centre as having improved, night-time safety was a concern for 62.5%; crime a concern for 49% and parking a concern for 27%. Day-time safety was ranked joint fourth (20%). The report concluded that:

> There are a number of deterrents to use of the city centre, but by far the most significant is concern about personal safety . . . Issues which are incorporated into a 'package' of concerns in this area include vandalism, begging, graffiti, the homeless, young people in groups in the street, car parking, subways and litter. (Beuret, 1995, section 4.4)

Crime figures and profiles for a city are double-edged; cities want to demonstrate a sufficient crime profile to warrant additional resources but equally do not want the publicity of the crime profile to be detrimental to the city's image or to deter people from coming to the city. Current figures suggest that there are approximately 250 public order incidents each month in Nottingham city centre; 80% are considered to be drink-related; less than 20% are reported and charges brought. Virtually all incidences occur on Friday or Saturday evening. In absolute terms these figures sound alarming, but when considered against the number of people using the city centre – up to 30,000 on each of Friday and Saturday evening – then the actual incidence could be considered to be relatively low (approximately one incident per thousand people per night). The police, however, do regard it as a problem. The incidences are also concentrated around pub closing time (after 11.00 pm) and after club closing time (after 2.00 am). Large numbers of people are all leaving pubs and clubs at the same time and trying to leave the city centre on inadequate public transport around the Market Square. The taxi and bus networks are strained; and many incidents occur as, for example, inebriated people jostle for taxis. People's fear of crime and consequent avoidance action, as discussed in Chapter 2, may have little relation to its actual incidence.

The city's experience during the 1996 European Football Championships suggest that staggered closing times will ameliorate this problem and proposals to stagger pub and club closing times to smooth out the demand on transport are being explored. It is argued that people have a fixed amount of money and capacity to consume alcohol and lengthening the opening hours will enable this to happen over a longer time period, thereby moderating its effects. To address public disorder, two other initiatives are currently being explored: the installation of a CCTV system in public places and a new byelaw to prohibit public consumption of alcohol.

CCTV

Nottingham has considered the installation of CCTV cameras since the early 1990s, and a traffic control surveillance system has been in operation since the mid-1970s with its control station in a bunker beneath the Trinity Square car park. In the early 1990s, a working group had been formed to discuss the issue and a number of options formulated. More recently, Nottingham has had a successful bid for Home Office funding in the second round of the CCTV Challenge initiative, receiving £200,000. The system is not seen as a panacea but will be linked in with other initiatives, such as, the Radio Shop Link and raising general awareness among security staff of crime problems.

There are already a number of CCTV systems in Nottingham (both major shopping centres, plus Debenhams and Marks & Spencer have systems). The argument accepted by the Home Office was that the existing private systems were the equivalent of matching funding. These systems are considered as Phase One. The Home Office will then fund Phase Two involving the installation of twelve state-of-the-art cameras on poles in and around the Market Square and on the major junctions of streets leading from the Market Square. The existing systems were installed to deal with retail crime, the new system will concentrate on public order issues. The existing CCTV systems in the city's council-owned car parks will be integrated into the system plus, it is hoped, the existing traffic management CCTV system in Nottingham currently operated by the county council – particularly if the new system can use the existing control bunker. Once the monitoring infrastructure is in place, it is relatively inexpensive to add further cameras to the network. It is, however, likely that if the city had further money to spend on CCTV, it would prioritise housing areas rather than the city centre. The system is likely to be up and running by September 1997, and, while linked to the police, will not be a police-operated system.

As a requirement of Home Office funding, the city council must establish a code of practice for the use of the CCTV system, its operators and the use of any tapes. This should provide some protection against possible infringements of civil liberties. In other cities, such as Northampton, it has been noted that CCTV systems tended to displace winos and beggars; while their welfare is a consideration, what should also be considered is the contribution of such people to the negative perceptions of city centres by other citizens as shown in the Perceptions Study.

Byelaws

Nottingham is also considering bringing in a byelaw to prohibit the consumption of alcohol on its city centre streets. Research for the Nottingham Crime Audit 'estimated that alcohol was a factor in 88% of incidents of criminal damage and 78% of all assaults' (KPMG Peat-Marwick, 1991). The city is about to start the required six month monitoring period needed to justify to the Home Office that a byelaw is required. Given the positive experience in other cities, such as Coventry, it is felt that the Home Office has become more permissive in granting approvals to create byelaws.

Public drinking has only become an explicit problem in Nottingham over the last two years. Arguably the city has become a 'victim of its own success' in promoting a vibrant night life. The number of licensed premises (particularly wine bars and cafes) and A3 uses (leisure centres, gyms and fitness clubs, etc., that also have bars) in Nottingham city centre has increased significantly over the last few years. The offence in the byelaw is likely to be continuing drinking after having been asked to desist. One of the more technical problems for the byelaw is to adequately define the area where the byelaw applies. Although the inner ring-road, for example, is a less distinct boundary than at Coventry, it is currently considered to be the most practicable area to exercise the byelaw. A less expansive area could simply lead to displacement to other parts of the city centre.

There are two major issues that the byelaw will need to address. The first is in dealing with winos. In terms of welfare provision, there is a 'wet centre' in Sneinton and a night shelter on Canal Street. Experience from other cities, suggests that the winos are likely to be displaced and it could be that they are displaced to nearby residential areas. In Coventry, it is recognised that there are still certain areas in the city centre where a 'blind eye' is turned to winos as – literally – they have nowhere else to go. The problem is therefore effectively 'contained' within a known area.

The second problem is enforcing the byelaw. There is a culture of 'circuit drinking' in Nottingham: rather than staying in a single pub all evening, groups of people move around and visit several pubs over the evening, often carrying their drinks with them. The problem is less one of broken glass and more of bottles and glasses being abandoned around the city centre on window ledges and in shop fronts once the drink has been consumed. To reduce the consequences of glasses being used as weapons (i.e., an example of the situational crime prevention technique of controlling the facilitators of crime) drinks could be served in plastic glasses, but many pub managers are reluctant to do this as it adversely affects custom. Furthermore, the trend is for many drinks to be served in bottles.

It is debatable whether the police currently have the available resources to enforce the byelaw and they could be overwhelmed. It is argued that the club and pub managers will need to take greater responsibility to ensure that there is sensible drinking and that drinks are not taken out of their premises and controlled outside areas. The city centre manager is also holding discussion with the major breweries – who have a vested interest in the city centre as one of their key markets – about joint initiatives to deal with problems of the misuse of alcohol.

CONCLUSION

The two city centres are very different types of public realm providing different functions as well as markets. Coventry is an example of Modernist design which was guided by safety concerns resulting from pedestrian-vehicular conflict and, hence, was created as a pedestrian precinct. It has design faults since, in the period when it was designed and built, perceptions of personal safety were not a social concern. Hence it has many 'nooks and crannies' and a multitude of obstructions to clear surveillance – both natural and CCTV – which create problems. Being a monofunctional shopping precinct owned by the city council, with a city centre bounded by an unmistakable ring-road, has enabled the city to define, isolate and bring about reasonably quick solutions to its problems. Its car park improvements and CCTV coverage have resulted in significant improvements. The high-profile alcohol ban seems to have worked and has gained public support.

Nottingham has both more complex problems and more opportunities than Coventry. In Nottingham, as in many towns, land ownership is highly fragmented. The multiplicity of ownership throughout the centre adds to its complexity and the challenges it has to address. As Beck & Willis (1995, p.149) observe, the position of the town centre manager when compared with out-of-centre shopping complex managers is very different:

> he or she has to deal with multiple ownership and tenancies, as well as the juxtaposition of private property and public-places. This leads to a complex web of loyalties and affiliations, compounded by a three-way obligation to serve the needs of the retailers and businesses, the local authority and the wider community of town-centre users . . . This may look like a 'basket' of valuable resources, but it can also be seen as a 'porridge' of competing and conflicting interests.

Nottingham has an attractive, mixed use city centre which has evolved over many years. It is both a successful retail centre and night-time entertainment centre. It is not clearly defined by a ring-road and even where the ring-road is clear, as at Maid Marion Way, the leisure uses cross over to the west side. Nottingham also has pockets of residential development, such as the Park Estate to the west on the edge of the city centre and increasingly the historic Lace Market within the city centre.

By force of circumstance, Coventry's recent history has been towards a form of 'revolutionary' change as a consequence initially of its destruction in the Second World War. By contrast, Nottingham has generally been more conservative and evolutionary with well-established partnerships and a sense of 'civic responsibility' among local businesses. Starting from a position of strength and well-thought out strategies, Nottingham has been able to take a subtler approach than Coventry's more 'law and order' stance. Environmental improvements including the car parks and management of the area, have paid dividends and Nottingham seems to be holding its position as a regional centre in the face of strong competition. It is significant however that it is now installing a CCTV system and considering a public drinking byelaw. In some ways, this predominantly night-time concern is a product of its own success.

Having significantly different structures, functions and hence differences in their public realms, the cities have adopted very different approaches. Both cities have generally achieved the goals they set. Coventry's high profile interventions may have raised concerns in the liberal establishment and press but seem to have strong support within Coventry. Unlike Los Angeles which created a centre for its elite office workers, Coventry has created a much safer shopping environment for virtually all of its citizens. Nottingham's 'softly, softly' approach, focusing on promoting positive images of the city centre, also seems to be working, although it still has some distance to go in terms of making the city centre safe for a wider community at night.

14

Towards Safer City Centres

This book has concentrated principally on the physical and managerial measures available to planners, urban designers, city centre managers and others concerned with safety in the public realm of city centres. As discussed in Chapter 1, it is debatable whether city centres will continue to exist in the way that we know them today. Out-of-centre regional shopping complexes are in fierce competition with city centres for retail and increasingly for leisure trade, while the transformation of previously centralised transport networks into grids has removed the central city's primacy within the network, heralding the emergence of Garreau's edge city and Fishman's perimeter cities. In the UK, however, current planning and urban policy is encouraging vital and viable town centres as sustainable urban developments. The UK Government would also like a significant proportion of the projected need for 4.4 million more homes to go into the city centres. For this to happen, city centres need to be attractive and safe.

Perhaps the strongest argument for the revitalisation of city centres is that people feel that something is missing in the placelessness of suburbia and the edge city, in shopping complexes and theme parks. Thus, they crave the excitement, variety, historicity, spectacle and carnival of real cities. Lovatt & O'Connor (1995, p.127), for example, observe how 'The 1980s saw the re-emergence of a concern with city centres as focal points for, and as symbolic of, a specifically urban way of life seemingly eroded in the 1970s'. Similarly, Robins (1996, p.46) describes a project working for the 'renaissance' of urban culture, while Rogers (1992, p.xvi) argues that:

> urban density provides the best setting for the easy, face-to-face interaction and communication that generates the scientific, technological, financial and cultural creativity that is the engine of economic prosperity in the post-industrial age.

There will be no revitalisation nor renaissance, however, unless urban areas and city centres are regarded as safe. Increased safety is a necessary – but not a sufficient – condition of revitalisation. Although it may not be possible to make city centres completely safe, nonetheless it is possible to make them safer so that more people perceive them to be safe and use them safely and comfortably. Their use by increasing numbers of people would lead to a revival of the public realm. The decline would be reversed and more people would discover – or re-discover – the rewarding experience of the urban public realm. The competing social arenas – most prominently, shopping malls – are not truly public spaces. This chapter will discuss the implications for the public realm as well as the consequences of the erosion of the public realm.

THE PROBLEM OF PUBLIC PLACES

As city centres represent perhaps the last significant concentrations of public space, their diminishing significance as arenas of public life has raised concerns about the decline of the so-called democratic public realm – democratic meaning, in this instance, universally accessible – and of public life more generally.

The public realm and public places generally are common property resources – resources to which there do not exist private property rights. Since the benefits can not be appropriated by an individual and are available to be shared by all those who have access, it does not necessarily reward any individual to make investments in improving or conserving the resource. As a result the public realm often becomes 'someone else's responsibility'. In the city centre, the public realm provides the scene and structure for private transactions in terms of retail, leisure and other services for the public. The decline of the public realm means a decline – and, in certain cases, the erosion – of these transactions. In recent years, the solution to the problem of the burden and upkeep of public space has often been to create proprietorship and 'ownership'. As the quality of the public realm and the private transactions that occur within it are so intertwined it is natural that such partnerships emerge and develop.

Those who manage the space seek to gain some return for their investment in its management in the form of consumption by its users. While in public space citizens have rights, in most forms of private space their presence is valued only in terms of their ability to consume. As Hayward & McGlynn (1995, p.322) argue: 'Consuming – retail goods and services – [has become] the be-all-and-end-all of all acceptable public assembly'. In quasi-public space, citizens are therefore reduced to consumers and their status as citizens accordingly diminished. If they are not sufficiently affluent to be consumers or if their presence will deter other consumers, then their presence may be unwanted.

THE PUBLIC REALM

The civic function of public space – the use of public space over and above consumption – is associated with the concept of the public realm. Cities are increasingly recognised as centres of social exchange, transactions and interaction between people. It is this social dimension of cities beyond their merely economic and commercial dimensions that is of interest in discussions of the public realm. In the contemporary period the significance of the public realm relative to the private realm has declined. Sennett (1986) refers to this as 'the fall of public man'. People have increasingly retreated into their private realms but, as Arendt (1958) contends, 'the shelter of the private' only reinforces one's own position, leading to loneliness and isolation. Loukaitou-Sideris (1996, p.100) argues that:

> The fragmentation of the public realm has been accompanied by fear, suspicion, tension and conflict between different social groups. This fear results in the spatial segregation of activities in terms of class, ethnicity, race, age, type of occupation and the designation of certain locales that are only appropriate for certain persons and uses (skid row park, ghetto, corporate plaza).

If the public realm is in decline, it is useful to consider what is being lost. The concept of the 'public' has interested many writers. Jurgen Habermas (1989), for example, saw the importance of a 'public sphere' as both a check and a balance on the state and the market. To Habermas, it is a space in which citizens have freedom of assembly and association and can confer in unrestricted fashion and express their opinions about matters of general interest. It is important to note however that Habermas' public sphere is not predicated upon the need for public space. Arendt (1954) is often cited to support ideas of a common interactive public realm. Ellin (1996, p.126), for example, argues that for Arendt the public realm satisfies three criteria: first, by outlasting mortal lives, it memorialises and thereby conveys a sense of history and society to individuals; secondly, it is established collectively and is an arena for diverse groups of people to engage in dialogue, debate and oppositional struggles; thirdly, it is accessible to and used by all. For Worpole (1992, p.5) the public space of the city centre has a simpler function and value as:

> important neutral territory, a site where people can mix and mingle without feeling socially embarrassed, where to some degree everybody is equal . . . the majority of people still feel that the town centre belongs to everyone.

It is, nevertheless, valid to question whether the loss of truly public space and of the public realm is a concern for people at large – whose individual market choices in the US and increasingly in the UK seem to favour the out-of-centre shopping malls and the edge city generally. The loss of a vibrant public realm may at present only be a concern for liberal academics who appreciate the subtle distinction between the public space of the city centre and the shopping mall's quasi-public realm (see for example Barker, 1996). But, should it be allowed to decline further, it will increasingly become a concern for the whole of society.

THE PHYSICAL PUBLIC REALM

The public realm consists of two parts: the *physical* public realm – the physical space – and the *socio-cultural* public realm – those activities occurring within that space. Although the latter are inherently more important, the former may create the preconditions – may even be a necessary precondition – for the latter to occur. The public realm therefore requires a spatially appropriate setting. Chapter 1 noted the failure of Modernist urban space design. Among the first to produce a coherent critique of Modernist theories of urban space were Rowe & Koetter (1975; 1978).

Rowe & Koetter (1978, p.50–85) described the spatial predicament of the Modernist city as being one of 'objects' and 'texture'. Their object is the sculptural building standing freely in space, while the texture is the background, continuous matrix of built form that provides the definition of that space by establishing the street corridor or the wall to a square. What many individual Modernist buildings and developments lacked was a positive response to the external – and often historic – context. Where all buildings are self-referential objects in space, there is no texture to define that space. With an increasing aggregation of such developments, a city loses its spatial qualities and coherence, becoming a 'jumble' of competing or isolated monuments and small complexes of buildings surrounded by roads and a sea of car parking, lacking the connections and positive spaces in-between that make sense of urban forms. Safdie (1987, p.153) argues that a city's legibility depends on 'the public domain as the connective framework between individual buildings . . . it does not exist today. We are unable to connect buildings as part of the urban experience.'

The result is a new kind of space and the type of street lauded by, for example Jane Jacobs, often no longer exists either physically or socially. As Loukaitou-Sideris (1996, p.97) describes:

> The result in spatial terms is 'a patchwork quilt of private buildings and privately appropriated spaces' (Trancik, 1986), cut off from the rest of the city fabric. The inward orientation and fragmentation of the city's spaces is in strong conflict with urbanistic objectives for coherence, continuity, linking of districts and social goals of integration, justice and equal access to spaces and amenities for all citizens.

The resultant city centre is one in which the structuring and connecting role of the physical public realm is lost. Hence private transactions tend to occur independently rather than interdependently. A city centre, such as Los Angeles, structured along these lines can only be safe by being either a single fortress or a series of fortresses. The rich complexity of the urban realm which peoples a city centre and makes it safe is lost as the people desert its streets and squares, only driving into and out of their offices, and to the shops and the few other facilities they use. The challenge for contemporary urban design has therefore been to restore the spatial discipline of the street and square, the spatial continuity and legibility of the city, and the positive design of the spaces between buildings.

THE SOCIO-CULTURAL PUBLIC REALM

The public realm is also a socio-cultural construct. Not only is a spatially defined physical public realm required, but that public realm needs to be animated by people. Arguably, however, the social and civic functions that might occur within public spaces have eroded along with the physical definition of that space. As Ellin (1996, p.149) describes:

> Activities which once occurred in the public realm have either been abandoned (e.g., liberal discourse) or usurped by more private realms, as leisure,

> entertainment, gaining information, and consumption are increasingly satisfied at home with television or computer. Or if one leaves home, these activities often take place in the strictly controlled uni-functional settings of the shopping mall, theme park, or varients thereof.

Contemporary urban design is intimately concerned with the public realm and creating a sense of place and place making (**Figure 14.1**). The presence of people turns spaces into places, making them living, working, organic parts of the city. Furthermore, the presence and activity of people will attract other people. One idea that has been argued in this book is that peopled places are safer places. The DoE Circular, *Planning Out Crime* (1994, section 14) states:

> One of the main reasons people give for shunning town centres at night is fear about their security and safety; one of the main reasons for that fear is the fact that there are very few people about. Breaking that vicious circle is a key to bringing life back to town centres.

The key to peopled places is the spatial concentration of different uses, especially residential uses. Creating mixed use environments is no longer primarily a problem of land use restrictions. Post war functional zoning policies have increasingly been abandoned and many cities positively encourage mixed use developments. Planning policies are more flexible with 'central area uses' or similar descriptions permitting retail, commercial or residential uses where they can be satisfactorily provided (Coupland, 1995). More recently, the current draft revision of *PPG1: General Policy and Principles* (DoE, 1996b) prioritises mixed use as a key dimension of future urban development. There remain however problems with the entrenched attitudes of funding institutions who have an aversion to developments that require more than minimal management. They also want to be able to buy or sell property with as few complications as possible and therefore prefer a single use. The current problems are due to the reluctance of developers to provide them and the lack of positive financial incentives to encourage mixed uses.

For urban spaces to become animated by people, the public realm has to offer what people want and desire. Furthermore, it has to do so in an attractive and safe environment. Animation can be encouraged by planning. MacCormac (1983a), for example, discusses the osmotic properties of streets: the manner in which the activities within buildings are able to percolate through and infuse the street with life and activity. He notes that there are certain uses which have very little relation to people in the street, and others with which they are intimately involved; the sense of human presence and vitality within the urban spaces depends upon these relationships.

MacCormac (1983a) establishes a hierarchy of uses supportive of an animated public realm. This does not suggest that some uses are unnecessary or have no place within an urban area, merely that they should have less claim to frontage onto the street and therefore onto the physical public realm. The extremes in MacCormac's hierarchy are car parking which has little or no significance to the passer-by and street markets which offer an intense series of transactions between the seller and the public, the stall and the street. Between

Figure 14.1 Pioneer Courthouse Square. The conscious creation of public spaces has enhanced the public realm of Portland, Oregon. Portland is one of the few remaining US cities that still has a liveable city centre.

these two lies a range of uses. To ensure busy, livelier spaces, the more interactive uses must be adjacent to those spaces. For reasons of personal safety, car parks and parking places must be closely related to well-used pedestrian areas.

The challenge in creating a lively urban city centre – or activity corridor – is to ensure that the most interactive uses claim the appropriate street frontages. Planning policies, design codes and frameworks have a role to play here. In Dublin's Temple Bar, the 1992 Development Programme introduced a detailed mixed-use plan including the vertical zoning of land-uses. This policy concentrated on the socio-cultural public realm and encouraged active ground floor uses, such as retail, bars, clubs, galleries and other cultural facilities to help animate the streets and provide a boost to the evening economy and, therefore, the safety of the city centre. The control over the upper floors was more relaxed and allows for a variety of more 'passive' uses, such as residential and office accommodation. There is a similar policy in Denver's LoDo where a distribution of uses is designed to encourage pedestrian life to make the city centre a safer place through street life and vitality. In the US, design codes in Traditional Neighbourhood Developments (TNDs) and Transit-Oriented Developments (TODs) seek to achieve a similar effect. Other factors which affect the pedestrian-friendly nature of the city centre are its *permeability* – the ease with which a pedestrian can move safely around the city centre – and its *legibility* – the ease with which pedestrians can navigate around the city centre.

Montgomery (1995b, p.104) argues that the animation of the city centre public realm can also be stimulated through planned programmes of cultural animation. One of the origins of this was Nicolini's programme in Rome in the late 1970s. The animation involves programming events and spectacles to encourage people to visit, use and linger in urban places. Programmes usually involve a varied diet of events and activities, such as lunchtime concerts, art exhibitions, street theatre, live music and festivals, across a range of times and venues. As people visit the area to see what's going on, urban vitality is further stimulated and the public realm becomes animated by having more people on the streets and in cafés, etc.

Initially, at least, it is difficult to provide sufficient density and concentration of activity to enable the whole city centre to be peopled. The concentration of activity into 'corridors' which act as the connective tissue between public places can provide the necessary density of people in certain places to render them sufficiently safe. Activity corridors need to be good quality environments, with good lighting and natural surveillance and closely related to car parking and public transport. Such corridors can act as catalysts for the revival of the city centre as a whole. Similarly, revived quarters bringing more people into the city centre, including residents, can also play an important role in the revitalisation and animation of city centres.

INCREASING PARTICIPATION IN THE PUBLIC REALM

There are at least two areas where positive actions are necessary to increase the vitality and activity of city centres. First, there is a need to increase the available range of activities, and secondly, a need to increase and broaden the range of age, gender, social and ethnic groups using the city centre. The latter involves increasing access by addressing the factors that actively deter those groups from using the city centre, such as the fear of crime, the sense that they are dominated by particular social groups, and worries about travel to and from the city centre.

Providing More Facilities

The consequence of post-war planning policies that favoured functional zoning was to reduce city centres to merely shopping and office centres whose life ends at 6 pm and whose evening – and predominantly pub-based entertainment – begins after 8 pm. During the day, the problem is one of retaining sufficient retail outlets and commercial uses to ensure a vibrant and well-peopled city centre but also one of supplementing this with other uses. A more pertinent problem in many city centres is that there is not enough to do at night for a wide range of age, gender, social and ethnic groups.

Two mutually-reinforcing processes have happened concurrently. Avoidance of the city centre by many social groups has meant that the facilities used by those groups have ceased to be viable in that location and have either closed or moved elsewhere. Simultaneously, the growing domination and colonisation of the city centre by a particular age group has meant that the available range of facilities has tended to be focused increasingly on the needs of this group.

This desertion of the city centre by a wide range of people is a modern version of the processes of invasion and succession defined in urban ecology. The young invaded the city centres of the UK at night, while women and the middle-aged have deserted them. To reverse this situation the contemporary challenge is to find out what people want and need from the public realm and to provide it. If this happens, then the city's public places will be lively and animated. The Twenty-Four Hour City concept provides a useful mechanism to consider the city centre more positively and expansively. Inevitably there is a 'chicken and egg' situation as to which comes first: more people or more for them to do.

The milieu and presentation of evening activities can also be changed to make it more welcoming to a wider range of social groups. The UK's pub culture has traditionally been predominantly male and youth-oriented, with a 'culture of heavy drinking' taking place behind opaque glass windows with little communication with the street (Bianchini,1994, p.309). Nevertheless, many pubs are changing. Extended opening hours have helped create a shift towards a more continental ambience, while a new generation of proprietors try to attract more women (**Figure 14.2**).

As discussed in Chapter 11, the management of alcohol in city centres requires a judicious mixture of control and relaxation. Experience in cities

Figure 14.2 The UK's pub culture has traditionally been male and youth-dominated – a culture of heavy drinking behind frosted glass windows. Many pubs, however, are now changing to create an atmosphere that is more attractive to women including large clear windows at the front so that they can see in before entering. Rather than pool tables and televised sport, the new pubs offer newspapers and fresh flowers; they serve filter coffee, light snacks and a good selection of wines; air conditioning sweeps away the smell of stale cigarettes and beery men (Tredre, 1995).

pursuing Twenty-Four Hour City strategies suggests that a relaxation of licensing hours can be beneficial. What is also required is a responsible attitude by those consuming alcohol. Hope's 1985 study for the Home Office and the COMEDIA (1991) *Out of Hours* report argued that as well as situational measures – altering the character of pubs and clubs or bringing back mixed uses to city centres – the root causes of a culture of heavy drinking and violence need to be addressed. A city centre's evening economy does not have to consist only of cafés, bars and other alcohol-related facilities. It may include facilities such as cinemas and theatres, live music, libraries and bookshops, etc. To enable greater access to these, the provision of crèche and child minding facilities might also be considered.

Increasing Access

Increasing access involves addressing those factors that either actively deter or result in a failure to attract other groups. There is a need to increase access in two areas: physical and emotional. To address physical access, safer transport and parking is necessary. Safer public transport is necessary to reassure people that they can travel to and from the city centre safely. For those who travel by private car, safe parking is required. As discussed in Chapter 9, for many people on-street parking in the evening is the preferred alternative. People use their cars not only as a convenient mode of travel but also because it makes them feel safer.

The biggest hurdle to increased participation in the use of the city centre is the perceptions of a lack of safety and the fear of crime. Ultimately, fear of crime may often put an exaggerated perspective upon people's images of city centres. Davis (1990, p.224), for example, observes in Los Angeles that 'white middle-class imagination, absent from any firsthand knowledge of inner-city conditions, magnifies the perceived threat through a demonological lens'. Nevertheless, as discussed in Chapter 2, Garrofolo (1981, p.856) argues that fear of crime is functional to the extent that it encourages people to take appropriate precautions. It is dysfunctional when it exaggerates *or* underestimates the risk of victimisation.

The *Home Office Standing Conference Report on the Fear of Crime* (Home Office, 1989), as discussed in Chapter 6, defined a four stage spectrum of feelings about the risk of victimisation: complacency; awareness; worry; and terror. The first level is *complacency* which might entail a failure to take proper precautions. A second level is that of a healthy *concern* and *awareness* which might entail the adoption of realistic preventative and precautionary measures; whereby the management of crime risk would be successfully integrated into daily life. The third level is that of *worry* and *fear* where there is a preoccupation with harm and danger leading the individual to adopt perhaps unnecessary measures to protect themselves but which affect the quality and richness of life. The fourth level is that of *terror* leading to an obsession with crime and a total disruption of life. The tendency is for women to be at the third and fourth levels and for men – especially young men – to be at the first level.

To reduce fear of crime also requires that the city centre environment is in good condition, well-maintained and that physical incivilities are curbed. Signs of physical and social disorder, such as graffiti, litter, broken windows, vandalised public property, vomit and urine in shop doorways, drunks and beggars, signal an environment that is out of control, unpredictable and menacing (Painter, 1996, p.52). Other improvements such as improved lighting schemes (see Chapter 7) and the installation of CCTV (see Chapter 8) may also be necessary.

Efforts to make a city centre safer should focus on the needs of the most vulnerable. As discussed in Chapter 4, if the city centre is safer for them, then it is safer for everyone. Public places should be culturally pluralist and inclusive, rather than segregated and exclusive. The problem however is that to make it safer for some people may be contingent on the exclusion of others. In the UK, in the 1980s, to create safer environments for shoppers, there were pressures to disenfranchise young people with low spending power 'whose appearance and conduct did not conform to the moral codes of well-ordered consumption'. (Bianchini, 1990, p.5). Nevertheless, as shown in Chapter 13, Coventry was surprisingly reluctant to practise a formal exclusion of even 'known troublemakers' from the city centre shopping area.

One particular aspect that may deter many people from using city centres is the presence of 'people who cause anxiety'. In the US, there have been instances of more instrumental policies involving the positive displacement of indigents. During the 1996 Olympic Games, for example, the authorities in Atlanta are reported to have given the homeless one-way Greyhound tickets to remove them from the city for the period of the Games. Although such sweeps are usually justified on public health grounds, they often have a political purpose. In Los Angeles in 1987, for example, the police swept the streets prior to the Pope's visit, while in San Francisco and Atlanta there were 'street cleaning' sweeps prior to and during the national political conventions hosted by each city in the 1980s (Goetz, 1992, p.547). Such displacement policies may be appropriate short term situational measures but do not provide a long term solution to the wider social problem.

As exclusion is usually risk averse, the civil liberties of some individuals will be infringed and some individuals or groups unjustly stigmatised. In this respect it is important to consider on what grounds the exclusion is made: preconceptions, prejudices or 'gut feelings' about appearance; the impact of their presence on others, particularly if the others are more affluent consumers who may – as a result – be deterred; or exclusion as a result of *actual* – rather than the *expectation* of – unacceptable behaviour. In terms of freely accessible public space, perhaps only the latter is really justifiable. In his research at Coventry (1990, p.24), Ramsey noted one of the key issues was 'the striking of a balance between the individual's right to consume alcohol out in the open and the evident desire of a good many people to avoid encountering public drinkers'.

Although it is necessary to respect the rights of individuals and groups who may be unjustly stigmatised and excluded, there is nonetheless a duty to protect individuals from incivilities. While exclusion can be considered an infringement of civil liberties, civil liberties are a doubled-edged issue and it is important to question exactly whose liberties are being infringed. Access to a safer city centre should be extended to all sections of the community, and thus efforts must continually be made to broaden participation in the use of the city centre. The desire for a safer city is not limited to the affluent and more mobile citizens, who – in any case – have the choice to reject city centres. Large numbers of elderly and working-class citizens who do not have the choice to reject the city centre, also want a safer city centre. Research at the University of Nottingham clearly indicated that it is these citizens who are more demanding of a safe city centre. A similar attitude was recorded in Coventry, which serves a mostly working class population, where residents overwhelmingly approved of the alcohol ban.

THE MANAGEMENT AND REGULATION OF PUBLIC SPACE

Safer city centres for all needs greater management and/or regulation of the public realm. Although this inevitably involves issues of freedoms, there are no absolute freedoms. The holders of freedom bear the responsibility not only for their own actions but also to ensure that they do not infringe the freedoms of others. In the contemporary period, the previously unwritten rules that governed public behaviour increasingly need to be codified. As discussed in Chapter 3, the introduction of new rules or procedures can be important in order to remove ambiguity between acceptable and unacceptable conduct. Equally, regulations may be necessary to establish an explicit standard of acceptable behaviour or – as regulations are usually framed negatively – to establish an explicit standard of unacceptable behaviour.

Local authorities and others who have the responsibility to manage the public realm may therefore introduce sets of 'rules'. Although city centre managers or property owners may not be able to control access to city centre streets, it may be possible to regulate the behaviour allowed on them. When rule setting, controlling authorities must ensure popular support otherwise the rules can be oppressive and/or unpopular leading to the city centre becoming less – nor more – attractive. It is important, therefore, that the rules are introduced consensually as has been the case with Coventry's alcohol byelaw.

Explicit Policing

The agency with an explicit law enforcement function and responsibility for the maintenance of public order is the 'public' police. In the UK, the view expressed in official policy on crime prevention in the 1980s and 1990s has been that the police cannot do it all themselves and need the active cooperation of other statutory and voluntary agencies and of the public. Nevertheless, Box *et al.* (1988, p.353) note: 'given the statistical importance of incivilities and fear . . . a major component in older citizens' confidence in the police . . . is what are they doing about order on the streets'.

One possible management strategy is for the police to rigorously enforce the law – a policy of zero rather than maximum tolerance. For example, William

Bratton, the former commissioner of the New York Police Department (NYPD), took office in 1994 promising a crackdown on so-called 'quality of life' crimes such as public drunkenness, graffiti and panhandling. Bratton subscribed to Wilson & Kelling's 'broken windows' theory of policing. His unofficial motto was 'zero tolerance' (Katz & Cohen, 1996) leaving no room for ambiguity.

Bratton decided the way to tackle serious crime was to tackle petty crime. He believed that a failure to clamp down on minor offences led to an escalation in the seriousness of crime and that the first priority of law enforcement was to improve the quality of life by ridding the streets of nuisance: beggars, noise, vandals and 'squeegee merchants' (Gibbons, 1996, p.19). As Bratton (from Gibbons, 1996, p.19) describes:

> Something had to be done about the quality of life offences, they were causing fear. We had aggressive begging, noise, graffiti, public drinking and urination. They're not serious in themselves, but they raise fear levels of those who witness them and contribute to more serious crimes.

Although the link between the more minor and the more serious crimes is difficult to comprehend, the results are impressive[1] and the idea has attracted interest in Britain. Cleveland's Chief Constable stated:

> If we don't check yobbish behaviour, we create a licence to commit other crimes. Young people seem to feel that they can do what they want and nobody's restraining them. We're saying there are limits to what society tolerates. If we deal with minor crime, it reinforces the idea of how to behave. (Shaw, in Gibbons, 1996, p.21).

It is debatable, however, whether the police have the level of resources to enable them to maintain the intensity of a broken windows approach in the longer term and across wider areas. It may be that saturation policing and 'zero tolerance' is only a short term policy and may work principally by displacement. Nevertheless, it may also be true that actions in the short term may ease problems in the longer term.

Implicit Policing

The image of an expressly 'law and order' city centre, while attractive to some, may be adversely oppressive to others generating fear of crime and offering an unappealing ambience. Some would prefer to live in a – at least outwardly – more tolerant society. It is also important to note the sometimes subtle distinction between a police state and a policed state and that other methods of control may also be effective. As Jacobs (1961, p.41) wrote: 'No amount of police can enforce civilisation where the normal, causal enforcement of it has broken down'.

1 In the two years before Bratton quit in April 1996, crime levels plummeted. Serious crime fell by a third overall, murders by almost 50%, motor crime by 40%, robberies by 32% and burglaries by 24% (Gibbons, 1996, p.19).

In this respect, it is instructive to consider the controls that already operate in theme parks. Disney Productions, for example, are able to handle large crowds of visitors (100,000 people per day and upwards of thirty million a year) in an orderly fashion, whereby potential trouble is anticipated and prevented. Jeffery Katzenberg, head of Disney's movie division, suggests that Disney World should be thought of 'as a medium-sized city with a crime rate of zero' (from Sorkin, 1992, p.231). Sorkin (1992, p.231) notes that although 'the claim is hyperbole, the perception is not; the environment is virtually self-policing'.

At Disney World, the 'opportunities for disorder' are minimised in a variety of ways: first, by constant instruction and direction; secondly, by physical barriers which severely limit the choice of available actions; and, thirdly, by the surveillance of 'omnipresent employees who detect and rectify the slightest deviation' (Shearing & Stenning, 1985, p.419). The control strategies are embedded in both the *environmental design* and the *management* of the theme park. These typically have other functions overshadowing their control function; thus, for example, every Disney Productions employee 'while visibly and primarily engaged in other functions, is also engaged in the maintenance of order' (Shearing & Stenning, 1985, p.419). The overall effect is to embed the control functions into the 'woodwork' where their presence is unnoticed but their effects are ever-present. Shearing & Stenning argue that Disney Productions' power rests both in the physical coercion it can bring to bear – if and when it needs to – and in its capacity to induce co-operation by depriving visitors of a resource they value. As a consequence, control becomes consensual; order is presented as being in the visitors' interests, and therefore 'order maintenance is established as a voluntary activity which allows coercion to be keep to a minimum' (Shearing & Stenning, 1985, p.419).

The City Centre and the Theme Park

The comparison with the maintenance of order in theme parks begs a further question for city centres. Not withstanding certain distinctions – theme parks are not cities; visitors pay an entrance fee; visitors are tourists who are on holiday; visitors just pass through, they don't live there – it is important to consider why people feel safe there. By the inclusion of leisure facilities, shopping malls are already becoming more like theme parks, and it is likely that city centres will continue to take on more – rather than fewer – of the characteristics of both theme parks and shopping malls. Nevertheless, in his introduction to the book *Variations on a Theme Park* – subtitled *The New American City and the End of Public Space* – Michael Sorkin (1992, p.xv) notes that the essays presented

> describe an ill-wind blowing through our cities, an atmosphere that has the potential to irretrievably alter the character of cities as the pre-eminent sites of democracy and pleasure. The familiar spaces of traditional cities, the streets and squares, courtyards and parks, are our great scenes of the

> civic, visible and accessible, our binding agents. By describing the alternative [the theme park], this book pleads for a return to a more authentic urbanity, a city based on physical proximity and free movement and a sense that the city is our best expression of a desire for collectivity.'

The criticism of theme parks has largely been on three grounds: first, aesthetic; secondly, authenticity; and thirdly, that it is a private space presenting itself as a public space and cannot therefore truly represent the 'desire for collectivity'. Nevertheless, a negative evaluation is always contingent on whether a feasible better option exists, since – in practical rather than theoretical terms – although flawed it may be the best we have.

Although many parts of city centres are already becoming physically similar to theme parks (see for example Crawford 1992; Boyer 1992), the aesthetic criticism is a challenge for architects, planners and developers to improve the quality of designs. There is also a certain elitism in this criticism. Many cultural critics find it difficult to accept that many people actually seem to enjoy these spaces and argue that it must either be a product of false consciousness or manipulation.

Sorkin (1992, p.xv) argues that although theme parks present their 'happy regulated vision of pleasure', they do so 'by stripping troubled urbanity of its sting, of the presence of the poor, of crime, of dirt, of work'. Nevertheless, Fainstein (1994, p.232) observes how the popularity of many out-of-centre shopping complexes and revitalised urban areas 'seems to drive the cultural critics into paroxysms of annoyance as they attempt to show that people ought to be continually exposed to the realities of life at the lower depths'. The dismissal as inauthentic presupposes a previous time when there was a more authentic urbanity, but the suggestion that there was ever a 'previous golden age when urban life conformed more closely to the model of tolerant diversity' may simply be false and nostalgic (Fainstein, 1994, p.233).

The third criticism – that theme parks are not genuine public spaces – is the most pertinent and persuasive. They are, as Sorkin argues, pseudo-public spaces: 'In the "public" spaces of the theme park or the shopping mall, speech itself is restricted: there are no demonstrations in Disneyland' (Sorkin, 1992, p.xv). It is debatable therefore the extent to which the same regulation and control strategies can be transferred to public spaces. In the private realm of the theme park people exercise self-exclusion. Those who do not want to accept the necessary regimentation and regulation in order to enjoy the amenity in safety, do not go. The public realm on the other hand is 'everybody's turf' where some individuals and groups feel that they can behave as they like without regard for others.

Some cities are however beginning to learn from theme parks and translate subtle control strategies from the private to the public realm. They employ people to work as 'capable guardians' – sometimes calling them ambassadors or urban rangers – who have a 'control' function but it is neither highlighted nor their main function. Moreover they train other employees who interface with the public in the city centre in an ambassadorial role who, when called upon, can also act as capable guardians. A number of questions however remain. In city centres, how far is it possible to exercise the sanction of depriving visitors of resources which they value? To what extent should visitors trade the rights of citizenship for safety? Increasing numbers of citizens, however, seem to be in favour of more regulation and it may be that a greater degree of potential exclusion in the interests of the greater good of public safety and order are required.

TOWARDS SAFER CITIES

This book has not sought to present a formulaic solution to the problem of making city centres safer. In making the city centre a safer place – in reviving the public realm – there is no panacea, 'miracle cure' or easy answer; there are only difficult choices. In this book we have argued that measures can be taken to reduce the opportunities for certain forms of crime in the city centre. These measures generally involve the management and design of the immediate environment – the opportunity – in which crimes occur.

It is recognised that there are inevitably difficulties of translating general principles into detailed design prescriptions and that detailed practices are not easily transposed from one location to another. There are few universally applicable prescriptive recommendations because all strategies and approaches are contingent on local circumstances and opportunities. It is also important to recognise that design is an integrative process, by which it is the totality – rather than the perfection of any single part in isolation – that is important. Thus, as Clarke (1992, p.vii–viii) notes, although case studies may be a source of ideas, they will rarely show exactly what to do.

With the foregoing qualifications and as a summation of this book, the following are offered as general guidance principles towards the creation of safer city centres.

THEORY

- Fear of crime is not irrational; furthermore, as safety cannot be guaranteed, it is functional *but only* to the extent that it encourages people to take sensible precautions.
- Crime can for practical purposes be understood as a reasonably rational decision-making process involving the trade off of costs and benefits.
- Criminal activity involves three phases – initial decision, search and target. As the decision phase is less amenable to intervention, practical measures can more constructively be focused at obstructing the offender's search and target phases.

PROCESS

- Public order is not solely the responsibility of the police.
- Emphasise a participatory research and evaluation process, allowing the people most affected to both define and 'own' the problem and suggest and 'own' solutions.

- Work through partnerships of city centre stakeholders and cultivate a sense of ownership and responsibility for the public space of the city centre.

DESIGN

- Design for the more vulnerable, especially for the elderly and for women; if the city centre is safer for them, then it is safer for everyone.
- The focus of design improvements should be on reducing the opportunities for crime to occur and avoiding situations where people have no choice of options.
- Integrate physical improvements with community development.
- Assess both personal safety and security of property, but – where they conflict – prioritise personal safety.
- Do not design to give an impression of safety where real danger exists.
- In terms of design, it is the totality that is important not the perfection of any individual part in isolation.
- Concentrate a sufficient density/intensity of activity in corridors or quarters to make these areas safer.

SURVEILLANCE & PEOPLING

- Use natural surveillance as a primary aid towards crime prevention.
- Ensure there is natural surveillance of the full length of streets, supplement this with good quality lighting and, if necessary, CCTV.
- Encourage a mix of land uses to increase the level of density of activity and create peopled places.
- Provide well-used, well-lit corridors of activity, closely related to transport and parking provision, that will provide a sufficient density of people to render them safe.
- Encourage the (re-)establishment of sizeable non-ghetto-ised residential populations in city centres.

MANAGEMENT

- Show that someone cares by ensuring good maintenance not only to eradicate neglect but to prevent buildings and other amenities giving off signals of neglect.
- (Re-)introduce 'capable guardianship' and human presence to help and reassure citizens in public places; rather than anonymous security guards, reintroduce employees and encourage others who have a positive relationship with the place and are committed to it.
- Remember that the human presence of the 'friendly bobby on the beat' is reassuring to most citizens.
- Use positive (benign) deflection to create safer areas, but ensure that the solution of one crime problem does not create a more serious one elsewhere.
- Where necessary, make selective use of containment and fuse areas.

POSTSCRIPT: SOCIAL AND SITUATIONAL CRIME PREVENTION

Although this book has advocated the use of opportunity reduction measures as being the most fruitful for its intended audience, their contribution is limited. Furthermore, this book does not seek to argue that the various approaches discussed would provide totally safe environments. The evidence suggests that while poor design facilitates crime, it also shows that good design does not necessarily prevent it. As discussed previously, Bottoms (1990, p.7) notes that while some parents lock cupboards or drawers to prevent their children from helping themselves to loose cash, chocolates and so forth, others prefer to socialise their children not to steal even if the opportunities are available. Although, in this example, social and situational measures may be equivalent ways (in the short run) of achieving similar objectives, they will produce radically different environments. The latter method is innately superior but it involves the giving of trust and the acceptance of responsibility. Without that taking of responsibility, it inevitably fails. As well as situational and opportunity reduction measures, actions need to be taken to encourage the acceptance of (moral) responsibility. Much of this is due to individual conduct and is often beyond the scope of situational and management measures. What is therefore also needed is the curbing of social incivilities by better personal behaviour and conduct within the public space of the city centre. Rather than a culture of excuses for poor behaviour in public places, what is required is a culture of expectations for more civility; such civility can only be encouraged and learned. This applies especially to one particular social group (Pitt (1996), for example, provides a valuable introductory discussion to the politics and practice of youth justice, identifying welfare, justice and corporatist models).

Many will recognise the general truth in Murray's (1990b, p.22–23) view that: 'young males are essentially barbarians for whom marriage – meaning not just the wedding vows, but the act of taking responsibility for a wife and children – is an indispensable civilising force'. Furthermore, actions are necessary to mitigate the underlying motivation to offend. Hirschi (1969) identified four elements in the informal social control of delinquency: 'commitments', 'attachments', 'involvements' and 'beliefs'. As Felson (1986, p.121) states: 'Lacking commitment to the future, attachments to others, or conventional involvements and beliefs in the rules, an individual has no handle which can be grasped, and informal social control is impossible'. Thus, through socialisation – the forming of the social bond – society gains a 'handle' on individuals to prevent rule breaking. The problem is that such socialisation is very difficult to effect. Such actions are beyond the scope of this book. The control of crime – and thereby the creation of safer city centres – requires actions which address all of the dimensions of the criminal act. In the interim, architects, planners, urban designers and other environmental managers have a responsibility to explore strategies for making city centres safer.

Bibliography

Agnew, R.S. (1985), 'Neutralising the impact of crime', *Criminal Justice & Behaviour*, Vol 12, pp 221–239.
Alexander, C. (1965), 'A city is not a tree', *Architectural Forum*, April 1965, reprinted in *Design*, No 6. February, 1966, pp 46–55.
Arendt, H. (1958), *The Human Condition*, Chicago, University of Chicago Press.
Arnot, C. (1995), 'Alfresco café culture takes the heat off city streets', *The Observer*, April 16th, p 12.
Ashworth, G.J. & Tunbridge, J.E. (1990), *The Tourist-Historic City*, London: Belhaven Press.
Association of Chief Police Officers (ACPO) (1990), *Policy Statement*, July 1990, London: ACPO.
Association of Town Centre Managers (1994), *An Introduction to Town Centre Management*, London: ATCM.
Atkins, S. (1989), *Critical Paths: Designing for Secure Travel*, London: The Design Council.
Atkins, S., Husain, S. & Storey, A. (1991), *The Influence of Street Lighting on Crime and the Fear of Crime*, Crime Prevention Unit Series, Paper 28, London: Home Office.
Audit Commission (1993), *Helping with Enquiries: Tackling Crime Effectively*, London: Audit Commission.
Baldwin, J., Bottoms, A.E. & Walker, M.A. (1975), *The Urban Criminal*, London: Tavistock.
Barker, M., Geraghty, J., Webb, B. & Key, T. (1993), *The Prevention of Street Robbery*, Crime Prevention Unit Paper No. 44, London: HMSO.
Barker, P. (1996), 'The future is here and now: lots of happy smiling people tripping to the shopping mall. But does it work?', *The Guardian* 2, Tuesday, October 8th, pp 2–3.
Barlow, J. & Gann, D. (1993), *Offices into Flats*, York: Joseph Rowntree Foundation.
Barr, R. & Pease, K. (1990), 'Crime Displacement and Placement', in Tonry, M. & Morris, N. (editors) (1990), *Crime and Justice: A Review of Research Vol 12*, Chicago: University of Chicago Press.
Barr, R. & Pease, K. (1992), 'A place for every crime and every crime in its place: An alternative perspective on crime displacement', in Evans, D.J. *et al* (editors) (1992) op cit.
Beck, A. & Willis, A. (1995), *Crime and Security: Managing the Risk to Safe Shopping*, Leicester: Perpetuity Press.
Beck, P.E. (1995), 'A protecting eye towards mall safety', *Building Magazine*, April 1995.
Bell, P., Fisher, J., Baum, A. & Green, T. (1990), *Environmental Psychology*, London: Holt, Rhinehart & Winston Inc.
Bennett, T. (1992), 'Themes and variations of Neighbourhood Watch', in Evans, D.J. *et al.* (editors) (1992), pp 272–285.

Bennett, T. (1991) 'The effectiveness of a police-initiated fear-reducing strategy', *The British Journal of Criminology*, Vol 31 (1), pp 1–14.

Bennett, T. (1986), 'Situational crime prevention from the offenders' perspective', in Heal, K. & Laycock, G. (editors) (1986) op. cit.

Bennett, T. & Wright, R. (1984), *Burglars on Burglary: Prevention and the Offender*, Aldershot: Gower.

Beuret, K. (1994), Paper presented at the Planning for Safety in Towns conference, Institute of Planning Studies, University of Nottingham, December 1994.

Beyleveld, D. (1979), 'Deterrence research as a basis for deterrence policies', *The Howard Journal of Penology & Crime Prevention*, Vol 18 (3), pp 135–149.

Bianchini, F. (1995) 'Night cultures, night economies', *Planning Practice & Research*, Vol 10 (2), May pp 121–126.

Bianchini, F. (1994), 'Night cultures, night economies', *Town & Country Planning*, Vol 63 (11), pp 308–310.

Bianchini, F. (1990), 'The crisis of urban public life in Britain, *Planning Practice & Research*, Vol 5 (3), pp 4–8.

Bianchini, F., Fisher, M., Montgomery, J. & Worpole, K. (1988), *City Centres, City Cultures*, Manchester: Centre for Local Economic Strategies.

Bianchini, F. & Parkinson, M. (1993), *Cultural Policy and Urban Regeneration: the Western European Experience*, Manchester: Manchester University Press.

Bianchini, F. & Schwengel, H. (1991), 'Re-imagining the city', in Corner, J. & Harvey, S. (1991), *Enterprise and Heritage: Crosscurrents of National Culture*, London: Routledge, pp 212–234.

Birmingham City Council (1990), *Multi-storey car park design survey*, Birmingham: Birmingham City Council.

Blackman, N. (1991), 'Ram-raiding – a retail crisis', *Security Industry*, December, pp 27–28.

Boggs, S.L. (1965), 'Urban crime patterns', *American Sociological Review*, Vol 30, pp 899–908.

Bottomley, A.K. & Pease, K. (1986), *Crime and Punishment: Interpreting the Data*, Milton Keynes: Open University Press.

Bottoms, E. & Wiles, P. (1992), 'Explanations of Crime and Place', from Muncie, J. *et al* (editors) (1966), pp 99–114 op. cit.

Bottoms, A.E. (1990), 'Crime prevention in the 1990s', *Policing and Society*, Vol 1, pp 3–22.

Bottoms, A.E. (1974), 'Review of defensible space by Newman, O.', *British Journal of Criminology*, Vol 14 (2), pp 203–206.

Bowlby, S. (1990), 'Women and the designed environment', *Built Environment*, Vol 16 (4), pp 245–248.

Box, S., Hale, C. & Andrews, G. (1988), 'Explaining fear of crime', *British Journal of Criminology*, Vol 28 (3), pp 340–356.

Boyer, M.C. (1994), *The City of Collective Memory: Its Historical Imagery and Architectural Entertainments*, Cambridge (Mass.): MIT Press.

Boyer, M.C. (1992), 'Cities for Sale: Merchandising History at South Street Seaport', from Sorkin, M. (editor) (1992), pp 181–204.

Brantingham, P.J. & Brantingham, P.L. (1993), 'Nodes, paths and edges: considerations on the complexity of crime and the physical environment', *Journal of Environmental Psychology*, Vol 13, pp 3–28.

Brantingham, P.J. & Brantingham, P.L. (editors) (1981; 2nd edition, 1991), *Environmental Criminology*, Beverly Hills, Ca.: Sage.

Brantingham, P.L. (1989), 'Crime prevention: The North American experience, in Evans, D.J. & Herbert, D.T. (1989), pp. 331–360 op. cit.

Brearley, N., Francis, P. & Matthews, R. (1993), *Nottingham City Centre Management: Community Safety Action Plan*, Leicester: University of Leicester Centre for the Study of Public Order.

Breheny, M. (1993), 'Planning the sustainable city region', *Town and Country Planning*, April.

Bridgeman, C. & Sampson, A. (1994), *Wise after the Event: Tackling Repeat Victimisation*, Report by the National Board for Crime Prevention, London: Home Office Police Research Group.

British Retail Consortium (1995), *Retail Crime Cost Survey*, London: BRC.

Brown, B. (1995), *CCTV In Town Centres: Three Case Studies*, Crime Detection and Prevention Series, Paper No 68, London: Home Office Police Research Group.

Buchanan, C. *et al*, (1963), *Traffic in Towns*, London: Ministry of Transport.

Buchanan, P. (1988), 'What city? A plea for a place in the public realm', *Architectural Review*, Vol 184 No 1101, November, pp 31–41.

Buckwalter, J. (1987), 'Securing shopping centres for inner cities', *Urban Land Magazine*, April 1987, pp 22–25.

Bulos, M. (1995), 'CCTV surveillance: safety or control?' Paper given at the Annual Conference of the British Sociological Association, Leicester, 10–13th April.

Butler, A.J.P. (1992), *Police Management* (2nd edition), Aldershot: Dartmouth Publications.

Calhoun, C. (1992) (editor), *Habermas and the Public Sphere*, Cambridge, Mass.: MIT Press.

Campbell, D. (1996), 'What is this thing called rage?', *The Guardian*, 6th January.

Campbell, D. (1995), 'Spy cameras become part of the landscape', *The Guardian*, 30th January, p 6.

Canter, D *et al* (editors) (1988), *Environmental Social Psychology*, Dordrecht: Nato ASI Series: Kluwer Academic Publishers.

Cardiff City Council (1995), *Cardiff City Centre: Prospectus for Investment*, Cardiff: Cardiff City Council.

Carmona, M. (1996), 'Controlling urban design – Part 2: Realizing the potential', *Journal of Urban Design*, Vol 1 (2), pp 179–200.

Carr, J. (1990), 'Out-of-town shopping: is the revolution over?', Royal Society of Arts Symposium.

Castells, M. (1989), *The Informational City: Information Technology, Economic Restructuring and the Urban-Regional Process*, Oxford: Basil Blackwell.

Castells, M. & Hall, P. (1994), *Technopoles: The making of Twenty-First-Century Industrial Complexes*, London: Routledge.

Chermayeff, S. & Alexander, C. (1963), *Community and Privacy: Towards a New Architecture of Humanism*, New York: Doubleday.

CIBSE (1992), *Lighting Guide — The Outdoor Environment*, LG6, London: CIBSE.

Citizens' Crime Commission of New York City (CCC) (1985), *Downtown Safety, Security and Economic Development*, New York: Downtown Research & Development Centre.

City of Glasgow District Council Planning Department (CGDC) (1995), *Results of the Merchant City's Residents' Attitudes Survey*, Glasgow: City of Glasgow.

City of Glasgow District Council Planning Department (CGDC) (1994), *Merchant City: Policy and Development Framework*, Glasgow: City of Glasgow.

City of Glasgow District Council Planning Department (CGDC) (1992), *The Renewal of the Merchant City*, Glasgow: City of Glasgow.

City of Toronto Planning & Development Department (CTPDD) (1990a), *Green Places? Safer Places: A Forum on Planning Safer Parks for Women*, Toronto: CTPDD.

City of Toronto Planning & Development Department (CTPDD) (1990b), *City Plan '91: Part 10 Planning for a Safer City*, Toronto: CTPDD.

City of Toronto (1988), *The Safe City – Municipal Strategies for Preventing Public Violence against Women*, Toronto: City of Toronto.

Clarke, M.J. (1987), 'Citizenship, community, and the management of crime', *British Journal of Criminology*, Vol 17 (4), Autumn, pp 384–400.

Clarke, R.V.G. & Mayhew, P. (1980) (editors), *Designing Out Crime*, Home Office Research & Planning Unit, London: HMSO.

Clarke, R.V. & Weisbund, D. (1994), 'Diffusion of crime control benefits: Observations on the reverse of displacement', in Clarke, R.V. (editor) (1994), *Crime Prevention Studies*, Vol 2, pp 165–183.

Clarke, R.V.G. (1992) (editor), *Situational Crime Prevention: Successful Case Studies*, New York: Harrow & Heston.

Clarke, R.V.G. (1980), 'Situational crime prevention: theory and practice', *British Journal of Criminology*, Vol 20 (2), pp 136–147. Also in Muncie, J. *et al* (editors) (1996), pp 332–342 op. cit.

Clarke, R.V. & Cornish, D.B. (1985), 'Modelling offenders' decisions: A framework for research and policy', in Tonry, M. & Morris, N. (1985) (editors), *Crime & Justice*, Vol 6, Chicago: University of Chicago Press, pp 147–85.

Cohen, L.E. & Felson, M. (1979), 'Social change and crime rate trends: A routine activity approach', *American Journal of Sociology*, Vol. 44, pp 588–608.

Coleman, A. (1985), *Utopia on Trial: Vision and Reality in Planned Housing*, London: Shipman.

Collins, R.G., Waters, E.B. & Dotson, A.B. (1991), *America's Downtowns: Growth, Politics and Preservation*, Washington DC: The Preservation Press.

Commission for Racial Equality (1987), *Living in Terror*, London: CRE.

COMEDIA in association with the Calouste Gulbenkian Foundation (1991), *Out of Hours: A Study of Economic, Social and Cultural Life in Twelve Town Centres in the UK*, London: COMEDIA.

Cornish, D.B. & Clarke, R.V. (1987), 'Understanding crime displacement: an application of rational choice theory', *Criminology*, Vol 25, pp 933–947.

Cornish, D.B. & Clarke, R.V.G. (1986a), 'Situational evaluation, displacement of crime and rational choice theory', in Heal, K. & Laycock, G. (1986) (editors), pp 1–16 op. cit.

Cornish, D.B. & Clarke, R.V.G. (1986b), *The Reasoning Criminal: Rational Choice Perspectives on Offending*, New York: Springer-Verlag.

Coupland, A. (1995), 'Mix no match for property market', *Planning* No 1105, 10 February.

Crawford, M. (1992), 'The World in a Shopping Mall', from Sorkin, M. (editor) (1992), pp 3–30 op. cit.

Crewe, L. & Hall-Taylor, M. (1991), 'The restructuring of the Nottingham Lace Market: industrial relic or new urban model?', *East Midlands Geographer*, Vol 14, pp 14–30.

Crime Concern (1989), *Working in Partnership with Local Government, the Business Sector and the Police to Meet the Challenge of Crime: A Briefing Paper*, Swindon: Crime Concern.

Crime Concern (1995a), *National Transport Review of Personal Security Measures: Survey of Working Practices and Initiatives*, Report for the Department of Transport, Swansea: Crime Concern.

Crime Concern (1995b), *National Transport Review of Personal Security Measures: Case Studies*, Report for the Department of Transport, Swansea: Crime Concern.

Crime Concern (1995c), *Personal Security on Transport: Concluding Report*, Crime Concern/Department of Transport, London: HMSO.

Davidson, I. (1995), 'Do we need cities any more?', *Town Planning Review*, Vol 66 (1), pp iii–vi.

Davidson, N. (1981), *Crime and Environment*, London: Croom Helm.

Davidson, R.N. (1989), 'Micro environments of violence', in Evans, D.J. & Herbert, D.T. (1989) op. cit.

Davies, S. (1994), 'They've got an eye on you', *The Independent*, 2nd November.

Davies, S. (1995), 'Welcome home Big Brother', *Wired*, May, pp 58–62.

Davies, S. (1996), *Big Brother: Britain's Web of Surveillance*, London: Macmillan.

Davis, M. (1990), *City of Quartz: Excavating the Future in Los Angeles*, London: Verso.

Davis, M. (1992a), *Beyond Blade Runner: Urban Control – the Ecology of Fear*, Westfield, New Jersey: Open Magazine Pamphlet Series.

Davis, M. (1992b), 'Fortress Los Angeles: the militarization of urban space', from Sorkin, M. (editor) (1992), pp 154–180 op. cit.

Department of the Environment/Welsh Office (1994), *Planning Out Crime*, London: HMSO.

Department of the Environment/Welsh Office (1996a), *PPG 6: Town Centres and Retail Developments*, London: HMSO.

Department of the Environment/Welsh Office (1996b), *PPG 1: General Principles and Policy (Consultation draft)*, London: HMSO.

Department of the Environment (1988a), *PPG 6: Major Retail Developments*, January 1988, London: HMSO.

Department of the Environment (1988b), *Action for Cities: Building an Initiative*, London: HMSO.

Department of the Environment (1993), *PPG 6: Town Centres and Retail Developments*, July 1993, London: HMSO.

Department of Transport (1995), *Personal Security on Public Transport: Guidelines for Operators*, Department of Transport Mobility Unit, London: HMSO.

Department of Transport (1995), *Government Response to the Transport Select Committee Recommendations: Taxis and Private Hire Vehicles*, Command Paper 2715, London: HMSO.

Department of Transport (1985), *Transport Act*, HMSO Section 106.

Ditton, J. & Duffy, J. (1983), 'Bias in the newspaper reporting of crime', *British Journal of Criminology*, Vol 23 (2), pp 159–166.

Ditton, J., Nair, G. & Phillips, S. (1993), 'Crime in the dark: a case study of the relationship between street lighting and crime', in Jones, H. (1993), *Crime and the Urban Environment*, Aldershot: Avebury/Gower.

Ditton, J. & Nair, G. (1994), 'Throwing light on crime: a case study of the relationship between street lighting and crime', *The Security Journal*, Vol 5 (3), July, pp 125–132.

Ditton, J. & Nair, G. (1995), 'SOX versus SON', *Lighting Journal*, Vol 60 (2), April–May.

Dont, R.W., (1992), *Street Gangs*, Chicago: Gang Crimes Section, Chicago Police Department.

DuBow, F., McCabe, E. & Kaplan, G. (1979), *Reactions to Crime: A Critical Review of the Literature*, Washington DC: National Institute of Law Enforcement and Criminal Justice, US Department of Justice.

Dunstan, J.A.P. & Roberts, S.F. (1980), 'Ecology, delinquency and socioeconomic status', *British Journal of Criminology*, Vol 20, pp 329–343.

Edwards, P. & Tilley, N. (1994), *Closed Circuit TV: Looking Out for You*, London: Home Office.

Ekblom, P. (1988), *Getting the Best Out of Crime Analysis*, London: Home Office Crime Prevention Unit.

Ekblom, P. & Simon, F. (1988), *Crime and Racial Harassment in Asian-Run Small Shops*, Crime Prevention Unit Paper No 15, London: Home Office.

Ellin, N. (1996), *Postmodern Urbanism*, Oxford: Blackwells Publishers Ltd.

Elliott, N. (1989), *Streets Ahead*, London: Adam Smith Institute.

Evans, D.J. & Herbert, D.T. (1989), *The Geography of Crime*, London: Routledge.

Evans, D.J., Fyfe, N.F. & Herbert, D.T. (editors) (1992), *Crime, Policing and Place: Essays in Environmental Criminology*, London: Routledge.

Fainstein, S. (1994), *The City Builders: Property, Politics, and Planning in London and New York*, Oxford: Blackwell Publishers.

Falk, N. (1991), 'Revitalising the heart of the town centre', *Planning* No 939, 11th October, pp 16–17.

Falk, N. (1994), 'Time for town centre action before the rot sets in', *Planning* No 1069, 20th May, pp 8–9.

Federal Bureau of Investigation (FBI) (1991), *Crime Statistics in the United States of America*, Washington DC: FBI.

Felson, M. (1994), *Crime and Everyday Life: Insights and Implications for Society*, London: Pine Forge Press.

Felson, M. (1986a), 'Predicting crime at any point on the city map', in Figlio, R.M., Hakim, S. & Rengert, G.F. (1986) (editors), *Metropolitan Crime Patterns*, New York: Criminal Justice Press.

Felson, M. (1986b), 'Linking criminal choices, routine activities, informal control, and criminal outcomes', in Cornish, D.B. & Clarke, R.V.G. (1986b), *The Reasoning Criminal: Rational Choice Perspectives on Offending*, New York: Springer-Verlag, pp 119–128.

Felson, M. (1987), 'Routine activities and crime prevention in the developing metropolis', *Criminology*, Vol 25, pp 911–931.

Fieldhouse, P. (1995), 'Central decline', *Landscape Design*, No 230, May, pp 32.

Fieller, L. & Peters, J. (1991), *Planning Principles for Chicago's Central Avenue*, Chicago: City of Chicago.

Fisher, B.S. & Nasar, J.L. (1992), 'Fear of crime in relation to three exterior site features: prospect, refuge and escape', *Environment & Behaviour*, Vol 24 (1), pp 35–36.

Fisher, B.S. & Nasar, J. (1995), 'Fear spots in relation to micro-level physical clues: exploring the overlooked', *Journal of Research in Crime & Delinquency*, Vol 32, pp 214–239.

Fisher, M. & Owen, U. (editors) (1991), *Whose Cities?*, Harmondsworth: Penguin.

Fishman, R. (1987), *Bourgeois Utopias: The Rise and Fall of Suburbia*, New York: Basic Books.

Fleming, R. & Burrows, J. (1988), *The Case for Lighting as a Means of Preventing Crime*, Research Bulletin No 22, London: Home Office Research & Planning Unit.

Forester, D., Chatterton, M. & Pease, K. (1988), *The Kirkholt Burglary Prevention Project, Rochdale*, Crime Prevention Unit Paper 13, London: Home Office.

Francis, A. (1991), 'Private nights in the city centre', *Town & Country Planning*, Vol 60 (10), pp 302–303.

Frieden, B.J. & Sagalyn, L.B. (1989), *Downtown, Inc.: How America Rebuilds Cities*, Cambridge (Mass): MIT Press.

Gabor, T. (1981), 'The crime displacement hypothesis; an empirical examination', *Crime and Delinquency*, Vol 26, pp 390–404.

Gabor, T. (1990), 'Crime displacement and situational prevention: towards the development of some principles', *Canadian Journal of Criminology*, Vol 32, pp 41–74.

Gans, H. (1962), *The Urban Villagers: Group and Class in the Life of Italian-Americans*, New York: Free Press.

Garofalo, J. (1981), 'The fear of crime: causes and consequences', *Journal of Criminal Law & Criminology*, Vol 72, pp 839–857.

Garreau, J. (1991) *The Edge City: Life on the New Frontier*, New York: Doubleday.

Geake, E. (1993a), 'The electronic arm of the law', *New Scientist*, May 8th, pp 19–20.

Geake, E. (1993b), 'Tiny brother is watching you', *New Scientist*, May 8th, pp 21–23.

Gerard Eve (Chartered Surveyors) (1995), *Wither the High Street?*, London: Gerard Eve.

Gerard, L. (1994), 'Tensions of a city looking over its shoulder', *The Observer*, January 9th, p 3.

Gibberd, F. (1955), *Town Design*, (2nd edition), London: Architectural Press.

Gibbons, S. (1996), 'Reclaiming the streets', *Police Review*, 13th September, pp 19–21.

Gibson, C. & Wright, M. (1995), *Radio Links – Communities Linked Together with Two Way Radios and with the Police*, London: Home Office Police Research Group.

Goetz, E.G. (1992), 'Land use and homeless policy in Los Angeles', *International Journal of Urban and Regional Research*, Vol 16, pp 540–554.

Gold, R. (1970), *Urban Violence and Contemporary Defensible Cities*, American Institute of Planners Journal, Vol 36, no. 3, pp146–159.

Goldstein, H. (1990), *Problem-Oriented Policing*, New York: McGraw-Hill.

Goodchild, B. (1994), 'Housing design, urban form and sustainable development: reflections on the future residential landscape', *Town Planning Review*, Vol 65 (2), pp 143–158.

Gordon, M.T., Riger, S., LeBailly, R. & Heath, L. (1981), 'Crime, women and the quality of urban life', in Stimpson, C. (editor) (1981), *Women and the American City*, Chicago: University of Chicago Press.
Gordon, M.T. & Riger, S. (1989), *The Female Fear*, New York: Free Press.
Gottfredson, M. & Hirschi, T. (1990), *A General Theory of Crime*, Stanford, Ca.: Stanford University Press.
Graham, P. & Clarke, J. (1996), 'Dangerous places: crime and the city', in Muncie, J. & McLaughlin, E., (1996), pp 143–181 op. cit.
Graham, S., Brookes, J. and Heery, D. (1995) Towns on the television: closed circuit television surveillance in British towns and cities, Newcastle: Working Paper 50, Dept. of Town and Country Planning, University of Newcastle-upon-Tyne.
Graham, S. & Marvin, S. (1996), *Telecommunications and the City: Electronic Space, Urban Places*, London: Routledge.
Griffiths, C. (1995), *Tackling Fear of Crime: A Starter Kit*, Police Research Group, London: HMSO.
Guessoum-Benderbouz, Y. (1994), unpublished survey of women and safety issues at the University of Nottingham.
Habermas, J. (1962; 1989), *The Structural Transformation of the Public Sphere* (translated by Burger, T. & Lawrence, F.), Cambridge, Mass: MIT Press.
Hack, G. (1974), *Improving City Streets for Use at Night: The Norfolk Experiment*, Cambridge, Mass.: Massachusetts Institute of Technology.
Hartly, J.E. (1974), *Lighting Reinforces Crime Fight*, Pittsfield: Butterheim Publishing Corporation.
Harvey, D. (1989a), *The Condition of Postmodernity: An Enquiry into the Origins of Cultural Change*, Oxford: Basil Blackwell.
Harvey, D. (1989b), *The Urban Experience*, Oxford: Basil Blackwell.
Hassington, J. (1985), 'Fear of crime in public environments', *Journal of Architectural & Planning Research*, Vol 2 pp 289–300.
Hayward, R. & McGlynn, A. (1995), 'The town centres we deserve?', *Town Planning Review*, Vol 66 (3), pp 321–328.
Heal, K. (1992), 'Changing perspectives on crime prevention: The role of information and structure', in Evans, D.J. *et al* (editors) (1992), pp 257–271 op. cit.
Heal, K. & Laycock, G. (editors) (1986), *Situational Crime Prevention: From Theory into Practice*, Home Office Research & Planning Unit, London: HMSO.
Healey, P. & Nabarro, R. (editors) (1990), *Land and Property Development in a Changing Context*, Aldershot: Gower.
Healey, P., Davoudi, S., O'Toole, M., Tavanoglus, & Usher, D. (editors) (1992), *Rebuilding the City: Property-Led Urban Regeneration*, London: E & F N Spon.
Healey, P., Cameron, S., Davoudi, S., Graham, S., & Madanipour, A. (editors) (1996), *Managing Cities: The New Urban Context*, London: John Wiley and Sons.
Heath, L. (1984), 'Impact of newspaper crime reports on fear of crime', *Journal of Personality and Social Psychology*, Vol 47, pp 236–276.
Heidenson, F. (1989), *Crime and Society*, London: Macmillan Press.
Herbert, D.T. (1982), *The Geography of Urban Crime*, London: Longman.
Herbert, D.T. & Davidson, N. (1994), 'Modifying the built environment: the impact of improved street lighting', *Geoforum*, Vol 25 (3), pp 339–350.
Herbert, D.T. & Hyde, S.W. (1985), 'Environmental criminology: testing some area hypotheses', *Transactions, Institute of British Geographers*, Vol 10, pp 259–274.
Hetherington, P. & Travis, A. (1994), 'Pounding to a brand new beat', *The Guardian Outlook*, 16–17 April, p 23.
Hiley, J. (1995), *Safer Bus Routes: Final Report*, Nottingham: Nottingham City Council.
Hill, D.M. (1994), *Citizens and Cities: Urban Policy in the 1990s*, London: Harvester Wheatsheaf.
Hillier, B. (1973), 'In defence of space', *RIBA Journal*, November 1973, pp 539–544.
Hillier, B. (1988), 'Against enclosure', in Teymur, N. & Wooley, T. (1988), *Rehumanizing Housing*, London: Butterworths, pp 63–88.

Hillier, B. & Hanson, J. (1984), *The Social Logic of Space*, Cambridge: Cambridge University Press.

Hillier, B. *et al* (1978), 'Space syntax: a new urban perspective', *Architects Journal*, 30th November, pp 47–63.

Hirschi, T. (1969), *Cause of Delinquency*, Berkeley: University of California Press.

Hirschi, T. (1986), 'On the compatibility of rational choice and social control theories of crime', in Cornish, D.B. & Clarke, R.V.G. (1986b), pp 105–118 op. cit.

Holden, A. (1992), 'Lighting the night: technology, urban life and the evolution of street lighting', *Places*, Vol 8 (2).

Home Office (1982), *British Crime Survey*, London: Home Office.

Home Office (1984), *Crime Prevention*, Home Office Circular 113/84, London: Home Office.

Home Office (1985), *Criminal Careers of Those Born in 1953, 1958 and 1963*, Home Office Statistical Bulletin 7/1985, London: Home Office.

Home Office (1987), *British Crime Survey*, London: Home Office.

Home Office (1988), *Safer Cities Programme*, London: Home Office.

Home Office (1989), *Home Office Standing Conference Report on the Fear of Crime*, London: Home Office.

Home Office (1990), *Crime Prevention: The Success of the Partnership Approach*, London: Home Office.

Home Office (1992), *British Crime Survey*, London: Home Office.

Home Office (1996a), CCTV can crack crime into the next century, Press Release 007/96, 9 January, London: Home Office.

Home Office (1996b), Carlisle cameras will collar criminals, Press Release 036/96, 9 February, London: Home Office.

Home Office (1996c), Technology tips the balance on criminals, Press Release 142/96, 13 May, London: Home Office.

Home Office (1996d), Neighbourhood Watch fighting crime in the community, Press Release 160/96, London, Home Office.

Home Office (1996e), More winners switch on CCTV to stamp out crime, Press Release 186/96, 26 June, London: Home Office.

Home Office (1996f), Eastbourne winners switch on CCTV to stamp out crime, Press Release 200/96, 26 June, London: Home Office.

Home Office (1996g), Private security guards to be given licence to work, Press Release 253/96, 14 August, London: Home Office.

Home Office (1996h), CCTV camera cash, Press Release 258/96, London: Home Office.

Home Office (1996i), No boozing in public for under 18s, Press Release 272/96, London: Home Office.

Home Office (1995a), The best crime prevention schemes prove crime can be beaten, Press Release 227/95, London: Home Office.

Home Office (1995b), Preventing crime into the next century, Press Release 260/95, London: Home Office.

Home Office (1995c), Spotting, catching and punishing criminals with CCTV, Press Release 302/95, London: Home Office.

Home Office (and Central Office of Information) (1994), *Practical Ways to Crack Crime The Handbook (4th edition)*, London: Home Office.

Home Office Crime Prevention Unit (1989), *Safer Cities Progress Report: 1988/1989*, London: HMSO.

Home Office Crime Prevention Unit (1990), *Safer Cities Progress Report: 1989/1990*, London: HMSO.

Home Office Crime Prevention Unit (1991), *Safer Cities Progress Report: 1990/1991*, London: HMSO.

Home Office Crime Prevention Unit (1992), *Safer Cities Progress Report: 1991/1992*, London: HMSO.

Home Office Standing Conference on Crime Prevention (1991), *Safer Communities: The Local Delivery of Crime Prevention Through the Partnership Approach*, London: Home Office.

Home Office, Department of Education and Science, Department of Environment, Department of Health and Social Security, and Welsh Office (1984), *Crime Prevention*, Home Office Circular 8/1984, London: Home Office.

Honess, T. & Charman, E. (1992), *Closed Circuit Television in Public Places; Its Acceptability and Perceived Effectiveness*, Home Office Police Research Group, Crime Prevention Unit Series Paper 35, London: HMSO.

Hope, T. (1985), *Implementing Crime Prevention Measures*, Home Office Research Study No 86, London: HMSO.

Hope, T. & Shaw, M. (1988), *Communities and Crime Reduction*, Home Office Research & Planning Unit, HMSO, London.

Hough, M. & Lewis, H. (1989), 'Counting crime and analysing risks: findings from the British Crime Survey', in Evans, D.J. & Herbert, D.T. (1989), *The Geography of Crime*, London: Routledge.

Hough, J.M., Clarke, R.V.G. & Mayhew, P. (1980), 'Introduction', in Clarke, R.V.G. & Mayhew, P. (1980) (editors), pp 1–17 op. cit.

Houghton, G. (1992), *Car Theft in England and Wales: The Home Office Car Theft Index*, Crime Prevention Unit Series Paper 33, London: Home Office.

House of Commons Environment Committee (1994), *Shopping Centres and their Future* (Volume 1), Session 1993–1994, London: HMSO.

Howard, M. (1995), Key Objectives 1996/97, Annex A of letter to all Chief Constables, 14 December.

Humberside Police (1995), *Annual Report 1994/95*, Kingston-upon-Hull: Humberside Police.

Husain, S. (1988), *Neighbourhood Watch in England and Wales: A Locational Analysis*, Crime Prevention Unit Paper 12, London: Home Office.

Institute of Social Research (ISR) (1975), *Public Safety: Quality of Life in the Detroit Metropolitan Areas*, Ann Arbor: University of Michigan; Survey Research Centre.

International Downtown Association (1993), *Centralised Retail Management*, Washington D.C.: International Downtown Association.

Institute of Planning Studies (IPS) (now Department of Urban Planning) (1994), *Planning for Safety in Towns: Report of Conference*, University of Nottingham, December 1994.

Jacobs, J. (1961, 1984 edition), *The Death and Life of Great American Cities: The Failure of Town Planning*, London: Peregrine Books in association with Jonathan Cape.

Jacobs, A. & Appleyard, D. (1987), 'Towards an urban design manifesto', *Journal of the American Planning Association*, Vol 53 (1), pp 112–120.

Jeffery, C.R. (1971, 1977 2nd edition), *Crime Prevention Through Environmental Design*, Beverley Hills, CA.: Sage.

Jeffery, S.S. & Radford, J. (1984), 'Contributory negligence or being a woman? The car rapist case', in Scraton, P. & Gordon, P. (editors) (1984), *Causes for Concern*, Harmondsworth: Penguin.

Jenks, M., Burbon, E. & Williams, K. (editors) (1996), *The Compact City: A Sustainable Urban Form?*, London: E & F N Spon.

Johnson, E. & Payne, J. (1986), 'The decision to commit a crime: an information-processing analysis', in Cornish, D.B. & Clarke, R.V.G. (1986b), *The Reasoning Criminal: Rational Choice Perspectives on Offending*, New York: Springer-Verlag, pp 170–185.

Johnson, J. (1987), 'Bringing it all back home: Ingram Square, Glasgow', *Architects Journal*, 6th May, pp 39–51.

Johnson, J. (1989), 'Merchant revival', *Architects Journal*, 3rd May, pp 36–51.

Johnson-Marshall, P. (1966), *Rebuilding Cities*, Edinburgh: Constable Ltd.

Jones, B. (1989) (editor), *Political Issues in Britain Today* (3rd edition), Manchester: Manchester University Press.

Jones, R. (1996), 'Planning positively for town centres', *Town & Country Planning*, July/August, pp 206–208.

Jones, R.L. (1975), *Crime Reduction through Increased Illumination*, Washington D.C.: National Institute of Justice.
Jones, S. (1991), Planning for women's safety, unpublished MA dissertation at the University of Nottingham.
Jones, T., Young, J. & McLean, B. (1986), *The Islington Crime Survey*, London: Tavistock.
Junger, M. (1987), 'Women's experiences of sexual harassment', *British Journal of Criminology*, Vol 22 (4), pp 358–83.
Katz, I. (1996), 'Dead end streets', *The Guardian* 2, August 27th, p 2–3.
Katz, I. & Cohen, N. (1996), 'Can we heed Big Apple's message?', *The Observer*, 2nd June.
Katz, P. (1994), *The New Urbanism: Towards an Architecture of Community*, New York: McGraw-Hill.
Kinsey, R., Lea, J. & Young, J. (1986), *Losing the Fight against Crime*, Oxford: Blackwell.
Kitchin, H. (1994), *A Watching Brief: A Code of Practice for CCTV*, London: LGIU.
Klynveld Peat Marwick, *et al*, (1991), *Counting Out Crime: the Nottingham Crime Audit*, Nottingham: Nottingham City Council.
Koffka, K. (1935), *Principles of Gestalt Psychology*, London: Kegan Paul.
Kolb, D. (1990), *Postmodern Sophistications: Philosophy, Architecture, and Tradition*, Chicago: University of Chicago Press.
Knox, P. (1987), Urban Social Geography: An Introduction (2nd edition), London: Longman.
La Grange, R.L. & Ferraro, K.F. (1989), 'Assessing age and gender differences in perceived risk and fear of crime', *Criminology*, Vol 27 (4), pp 697–719.
La Grange, R.L., Ferraro, K.F. & Suponcic, M. (1992), 'Perceived risk and fear of crime: role of social and physical incivilities', *Journal of Research in Crime & Delinquency*, Vol 29 (3), p 311–334.
Landry, C. & Bianchini, F. (1995), *The Creative City*, London: Demos.
Lawrence, R.J. (1987), *Houses, Dwellings and Homes: Design, Theory, Research and Practice*, New York: Wiley.
Laycock, G. & Austin, C. (1992), 'Crime prevention in parking facilities', *Security Journal*, Vol 3, pp 154–159.
Laycock, G. & Heal, K. (1989), 'Crime prevention: the British Experience', in Evans, D.J. & Herbert, D.T. (1989), pp 315–330 op. cit.
Lea, J. & Young, J. (1984), 'Relative deprivation', in Muncie, J. *et al* (editors) (1996), pp 136–144 op. cit.
Lee, T. (1971), 'Psychology and architectural determinism (part one)', *Architects Journal*, Vol 154, 4th August, pp 253–262.
Leeds City Council (1995), *Leeds: 24 Hour City Licensing*, Legal Services Department, Leeds: Leeds City Council.
Leicestershire Constabulary and Leicester City Council (1993), *Crime Prevention by Planning Design*, Leicester City Council, Leicester: City Planning Department.
Lewis, D.A. & Maxfield, M.G. (1980), 'Fear in the neighbourhoods: an investigation of the impact of crime', *Journal of Research in Crime & Delinquency*, Vol 17, 160–189.
Lewis, E. (1995), *Truancy – The Partnership Approach*, Stoke-on-Trent: Smith Davis Press.
Liberty (1989), *Who's watching over you? Video surveillance in public places*, Briefing Paper No 16, London: Liberty.
Liddle, A.M. & Gelsthorpe, L.R. (1994a), *Inter-Agency Crime Prevention: Organising Local Delivery*, Police Research Group, Crime Prevention Unit Series Paper 52, London: Home Office.
Liddle, A.M. & Gelsthorpe, L.R. (1994b), *Crime Prevention and Inter-Agency Co-operation*, Police Research Group, Crime Prevention Unit Series Paper 53, London: Home Office.
Liddle, A.M. & Gelsthorpe, L.R. (1994c), *Inter-Agency Crime Prevention: Further Issues*, Police Research Group, Crime Prevention Unit Series Supplementary Paper to Papers 52 & 53, London: Home Office.

Local Government Information Unit (LGIU) (1994), *Candid Cameras – A Report on Closed Circuit Television*, London: LGIU.

Local Transport Today (1990), 'Easing women's travel fears; a suitable case for treatment?', *Local Transport Today*, 17 October, pp 3.

Lofland, L.H. (1989), 'The morality of urban public life: the emergence and continuation of a debate', *Places*, Vol 6 (3).

Lotz, R. (1979), 'Public anxiety about crime', *Pacific Sociological Review* 22, pp 241–54.

Loukaitou-Sideris, A. (1996), 'Cracks in the city: addressing the constraints and potentials of urban design', *Journal of Urban Design*, Vol 1 (1), pp 91–104.

Lovatt, A. (1994), *More Hours in the Day*, Manchester Institute for Popular Culture, Manchester: Manchester Metropolitan University.

Lovatt, A. (1993), 'Cultural identity through the evening economy', in *Transcript of the First National Conference on the Evening Economy*, Manchester: Manchester Institute for Popular Culture.

Lovatt, A. & O'Connor, J. (1995), 'Cities and the night time economy', *Planning Practice & Research*, Vol 10 (2), May, pp 127–134.

Lukes, S. (1985), *Marxism and Morality*, Oxford University Press, Oxford.

Lyon, D. (1994), *The Electronic Eye: The Rise of Surveillance Society*, London: Polity Press.

MacCormac, R. (1987), 'Fitting in offices', *Architectural Review*, May 1987, pp 62–67.

MacCormac, R. (1983a), 'Urban reform: MacCormac's manifesto', *Architects Journal*, 15th June, 1983, pp 59–72.

MacCormac, R. (1983b), 'The architect and tradition 2: tradition and transformation', *Royal Society of Arts Journal*, November 1983, pp 740–753.

Maltz, M.D., Gordon, A.C. & Friedman, W. (1990), *Mapping Crime in its Community Setting: Event, Geography, Analysis*, New York: Springer-Verlag.

Manchester City Council (1994a), *Report for Resolution, Policy & Resources Committee*, City Centre Sub-Committee, March 1994, Manchester: Manchester City Council.

Manchester City Council (1994b), *Report for Resolution, Policy & Resources Committee*, City Centre Sub-Committee, May 1994, Manchester: Manchester City Council.

Manchester City Council (1995), *City Development Guide: 1995*, Manchester: Manchester City Council.

Marsh, H.L. (1991), 'A comparative analysis of crime coverage in newspapers in the United States and other countries from 1960–1989: a review of the literature', *Journal of Criminal Justice*, Vol 19, pp 67–80.

Martinson, R. (1974), 'What works? Questions and answers about penal reform', *Public Interest*, No 35, p 25.

Marx, G.T. (1986), 'The iron fist and the velvet glove: totalitarian potentials within democratic societies', in Short, J. (1986) (editor), *The Social Fabric: Dimensions and Issues*, Beverly Hills, CA.: Sage.

MATRIX (editors) (1984), *Making Space: Women and the Man-Made Environment*, London: Pluto Press.

Mawby, R.I. (1977a), 'Kiosk vandalism: a Sheffield study', *British Journal of Criminology*, Vol 17 (1), pp 30–46.

Mawby, R.I. (1977b), 'Defensible space: a theoretical and empirical appraisal', *Urban Studies*, Vol 14, pp 169–179.

Mayhew, P., Maung, N.A. & Mirrlees-Black, C. (1993), The 1992 British Crime Survey, Research Planning Unit, London: Home Office.

Maxfield, M. (1987), *Explaining Fear of Crime: Evidence from the 1984 British Crime Survey*, Home Office Research & Planning Unit Paper No 43, London: HMSO.

Maxfield, M. (1984), *Fear of Crime in England and Wales*, London: HMSO.

Mayhew, P. (1979), 'Defensible space: the current status of a crime prevention theory', *The Howard Journal of Penology and Crime Prevention*, Vol 18, pp 150–159.

Mayhew, P. (1981), 'Crime in public view: surveillance and crime prevention', in Brantingham, P.J. & Brantingham, P.L. (editors) (1981; 2nd edition, 1991), pp 119–134.

Mazur, A. (1991), *Downtown Development – Chicago: 1989–1992*, Chicago: City of Chicago.
McKie, R. (1994), 'Never mind the quality, just feel the collar', *The Observer*, November 13th, pp 1.
McLaughlin, E. & Muncie, J., (editors) (1996), *Controlling Crime*, London: Sage Publications/The Open University.
McClintock, H. (1994), *Report of Workshop on Light Rail System, Planning for Safety in Towns: Report of Conference*, University of Nottingham Institute of Planning Studies, December 1994.
McCrystal, C. (1996), 'Blitz and starts', *The Observer Review*, April 28th, pp 1–2.
Meikle, J. (1996), ' "Have a nice day" plan to enhance urban living', *The Guardian*, April 20th 1996, p 6).
Merry, S. (1981), 'Defensible space undefended: social factors in crime prevention through environmental design', *Urban Affairs Quarterly*, 16, pp 397–422.
Miethe, T.D. & Meier, R.T. (1990), 'Opportunity, choice and criminal victimisation: a test of a theoretical model', *Journal of Research in Crime and Delinquency*, Vol 27, pp 243–266.
Michener, J.A. (1960), *The Quality of Life*, New York: Secker & Warburg.
Milder, N.D. (1987), Crime and downtown revitalisation', *Urban Land*, September, pp 16–25.
Ministry of Town & Country Planning (1947), *Advisory Handbook for the Redevelopment of Central Areas*, London: Ministry of Town & Country Planning.
Mirrlees-Black, C. & Ross, A. (1995), *Crime Against Retail Premises in 1993*, Research Findings No 26, London: Home Office Research and Statistics Department.
Montgomery, J. (1990), 'Cities and the art of cultural planning', *Planning Practice & Research*, Vol 5 (3), pp 9–16.
Montgomery, J. (1994), 'The evening economy of cities', *Town & Country Planning*, Vol 63 (11), pp 302–307.
Montgomery, J. (1995a), 'The story of Temple Bar: creating Dublin's cultural quarter', *Planning Practice & Research*, Vol 10 (2), pp 135–172.
Montgomery, J. (1995b), 'Urban vitality and the culture of cities', *Planning Practice & Research*, Vol 10 (2), pp 101–109.
Montgomery, J. (1995c), 'The evening economy of cities', in *Transcript of the First national Conference on the Evening Economy*, Manchester: Manchester Institute for Popular Culture.
Moughtin, J.C. (1992), *Urban Design: Street and Square*, Oxford: Butterworth Heineman.
Multan, G. & Wilkinson, H. (1995), 'Fast forward', *The Guardian 2*, June 6th, pp 1–2.
Muncie, J. (1996), 'The Construction and Deconstruction of Crime', in Muncie, J., & McLaughlin, E. (editors) (1996), pp 5–64 op. cit.
Muncie, J., & McLaughlin, E. (editors) (1996), *The Problem of Crime*, London: Sage Publications/The Open University.
Muncie, J., McLaughlin, E. & Langan, M. (editors) (1996), *Criminological Perspectives: A Reader*, London: Sage Publications/The Open University.
Murray, C. (1984), *Losing Ground: American Social Policy 1950–1980*, New York: Basic Books.
Murray, C. (1990a), 'How to win the war on drugs', *New Republic*, May 21st, 1990, pp 19–25.
Murray, C. (1990b), *The Emerging British Underclass*, London: The IEA Health & Welfare Unit.
Nair, G., Ditton, J. & Phillips, S. (1993), 'Environmental improvements and the fear of crime', *The British Journal of Criminology*, Vol 33 (4), pp 555–561.
Nasar, J. & Fisher, B. (1993), '"Hot spots" of fear and crime: a multi-method investigation', *Journal of Environmental Psychology*, Vol 13, pp 187–206.
National Association for Local Government Women's Committees (1991), *Responding with Authority – Local Authority Initiatives to Counter Violence Against Women*, London: NALGWC.

National Trust for Historic Preservation (1988), *Revitalising Downtown, 1976–1986*, Washington D.C.: US National Trust for Historic Preservation/Urban Institute.
Nelson, A. (1996), 'Security-bred insecurity', *Town & Country Planning*, July/August, pp 195–196.
Netherlands Ministry of Justice (1985), *Society and Crime: A Policy Plan for the Netherlands*, The Hague: Ministerie van Justitie.
Newman, O. (1973), *Defensible Space: People and Design in the Violent City*, London: Architectural Press.
Newman, O. (1980), *Community of Interest*, New York: Anchor Press/Doubleday.
Newman, O. (1995), 'Defensible space: a new physical planning tool for urban revitalisation', *Journal of the American Planning Association*, Vol 61 (2), Spring, pp 149–153.
Nicholson, L. (1995), *What Works in Situational Crime Prevention? A Literature Review*, The Scottish Office Central Research Unit Paper, Edinburgh: Scottish Office.
Nottingham City Council (1995), *Outdoor Cafés Design Brief*, Nottingham: Nottingham City Council.
Nottingham City Centre Management (1993), *CCTV Challenge Competition 1996/97 Bid*, Nottingham: Nottingham City Centre Management.
Nottingham City Centre Management (1993), *Improving the City Centre Experience: Business Plan 1993–97*, Nottingham: Nottingham City Centre Management.
Nottingham Evening Post (1992), 'Post comment: don't let yobs rule', *Evening Post*, September 25th, p 4.
Nottingham Safer Cities Project (1989), *Nottingham Safer Cities Project: Action Plan 1989/90*, Nottingham: Nottingham Safer Cities Project.
Nottingham Safer Cities Project (1990), *Steering Group Report on Safety in the City Centre*, Nottingham: Nottingham Safer Cities Project.
O'Connor, J. (1993), 'Towards the twenty-four hour city', in *Transcript of the First National Conference on the Evening Economy*, Manchester: Manchester Institute for Popular Culture.
O'Dea, W. (1958), *The Social History of Lighting*, London: Routledge & Kegan Paul.
Oc, T. (1991), 'Planning natural surveillance back into city centres', *Town & Country Planning*, September, pp 237–239.
Oc, T. & Trench, S. (1990) (editors), *Current Issues in Planning*, London: Gower.
Oc, T. & Trench, S. (1993), 'Planning and shopper security', in Bromley, R.D.F. & Thomas, C.J. (1993), *Retail Change*, London: UCL Press, pp 153–169.
Oc, T. & Trench, S. (editors) (1995), *Current Issues in Planning (Vol II)*, London: Gower-Avebury.
Osborn, F.J. & Whittick, J. (1977), *The New Towns*, London: Leonard Hill.
Osbourn, S. & Shaftoe, H. (1995), *Safer Neighbourhoods: Successes and Failures in Crime Prevention*, York: Safer Neighbourhood Unit.
Pain, R.H. (1991), 'Space, sexual violence and social control: integrating geographical and feminist analyses of women's fear of crime', *Progress in Human Geography*, Vol 15 (4), pp 415–431.
Pain, R.H. (1995), 'Elderly women and violent crime: the least likely victims?', *British Journal of Criminology*, Vol 35 (4), pp 584–598.
Painter, K. (1988), *Lighting and Crime Prevention: The Edmonton Project*, Middlesex Polytechnic: Centre for Criminology.
Painter, K. (1989), *Lighting and Crime Prevention for Community Safety: The Tower Hamlets Study: First Report*, Middlesex Polytechnic: Centre for Criminology.
Painter, K. (1993), 'Review of Atkins, Hussain & Storey (1991)', *British Journal of Criminology*, Vol 33, pp 139–141.
Painter, K. (1996), 'It's safety first and last', *Landscape Design*, March 1996, p 52.
Palmer, A. (1996), 'If they can't drive – will they still steal?', *The Sunday Telegraph*, October 13th, pp 28.
Park, R.E., Burgess, E.W. & MacKenzie, R.D. (1925), *The City*, Chicago: University of Chicago Press.
Paumier, C.B. (1988), *Designing the Successful Downtown*, Washington D.C.: Urban Land Institute.

Peel CPTED Committee (1994), *Crime Prevention Through Environmental Design Principles*, Peel, Ontario: Region of Peel/City of Brampton/Corporation of the Town of Caledon; Mississauga/Peel Regional Police/Ontario Provincial Police.

Petherick, A. & Fraser, R. (1992), *Living Over The Shop: A Handbook for Practitioners*, York: University of York.

Phillips, S. & Cochrane, R. (1988), *Crime and Nuisance in the Shopping Centre: A Case Study in Crime Prevention*, Crime Prevention Unit, Paper No 16, London: HMSO.

Phillips, T. (1995), 'Spies on the streets', *The Guardian 2*, 21st September.

Pisarski, A.E. (1987), *Commuting in America: A National Report on Commuting Patterns and Trends*, Westport, Conn.: ENO Foundation for Transportation.

Pitts, J. (1996), 'The politics and practice of youth justice', in McLaughlin, E. & Muncie, J., (editors) (1996), pp 249–291 op. cit.

Pocock, J. (1996), The High Street fights back, unpublished MA dissertation at the University of Nottingham.

Pollock, L. (1994), 'Scorned patrol', *The Guardian*, 6th April, p 12.

Poole, R. (1994a), *Safer Shopping II – The Identification of Opportunities for Crime and Disorder in Shopping Malls (2nd Volume)*, Birmingham: West Midlands Police with Police Research Group, London: Home Office.

Poole, R. (1994b), *Operation Columbus: Travels in North America – A Personal Journal*, Birmingham: West Midlands Police.

Poole, R. with Donovan, K. (1991), *Safer Shopping – The Identification of Opportunities for Crime and Disorder in Shopping Malls (1st Volume)*, Birmingham: West Midlands Police with Police Research Group, London: Home Office.

Poyner, B. (1992), 'Situational crime prevention in two parking facilities', in Clarke, R.V.G. (editor) (1992) op. cit.

Poyner, B. & Webb, B. (1991), *Crime Free Housing*, Oxford: Butterworth Architecture.

Poyner, B. (1986), 'A model for action', in Heal, K. & Laycock, G. (editors) (1986), pp 25–39 op. cit.

Poyner, B. (1983), *Design Against Crime: Beyond Defensible Space*, London: Butterworths.

Pressman, N. (1985), 'Forces for spatial change', in Brotchie, J. *et al*, (editors) (1985), *The Future of Urban Form*, London: Croom Helm, pp 349–61.

Preston, C. (1995), 'Someone's watching', *The Guardian Society*, 22nd March, pp 2–3.

Pritchard, S. (1995), 'Advent of the High Street evangelists', *Planning Week*, 7th July, pp 16–17.

Project for Public Spaces (PPS) (1984), *Managing Downtown Public Spaces*, Chicago: APA Publications.

Proshansky, H.M., Ittelson, W.H. & Rivlin, L.G. (editors) (1970), *Environmental Psychology: Man and his Physical Setting*, New York: Holt, Rinehart & Winston.

Punter, J. (1990), 'The privatisation of the public realm', *Planning Practice & Research*, Vol 5 (3), pp 17–21.

Rajgor, G. (1994), 'Bringing confidence back to car parks', *Public Service and Local Government*, August 1984, pp 19–20.

Ramsey, M. (1990), *Lagerland Lost? An Experiment in Keeping Drinkers off the Streets in Central Coventry and Elsewhere*, Crime Prevention Unit Paper 22, London: Home Office.

Ramsey, M. (1991), 'A British experiment in curbing incivilities and fear of crime', *Security Journal*, Vol 2, pp 120–125.

Ramsey, M. & Newton, R. (1994), *The Effect of Better Street Lighting on Crime and Fear: A Review*, Crime Prevention Unit, London: Home Office.

Rand, G. (1984), 'Crime and environment: a review of the literature and its implications for urban architecture and planning', *Journal of Architecture & Planning Research*, Vol 1 pp 3–20.

Reeve, A. (1996), 'The private realm of the managed town centre', *Urban Design International*, Vol 1 (1), pp 61–80.

Rengert, G.F. & Wasilchick, J. (1985), *Suburban Burglary*, Springfield, IL.: Chas C. Thomas.

Reppetto, T.A. (1976), 'Crime prevention and the displacement phenomenon', *Crime & Delinquency*, April, pp 166–77.

Rhodes, W.M. & Conley, C. (1981), 'Crime and mobility: an empirical study', in Brantingham, P.J. & Brantingham, P.L. (1981; 2nd edition, 1991), pp 167–188 op. cit.

Riger, S., Gordon, M.T. & Le Bailly, R.K. (1982), 'Coping with crime: women's use of precautionary behaviours', *American Journal of Community Psychology*, Vol 10, pp 369–386.

Robins, K. (1991), 'Tradition and translation: national culture in its global context', from Corner, J. & Harvey, S. (editors) (1991), *Enterprise and Heritage: Crosscurrents of National Culture*, London: Routledge, pp 21–44.

Robins, K. (1996), 'Collective emotion and urban culture', from Healey, P., Cameron, S., Davoudi, S., Graham, S., & Madanipour, A. (editors) (1996), *Managing Cities: The New Urban Context*, London: John Wiley & Sons, pp 45–62.

Rock, P. (1990), *Helping Victims of Crime*, Oxford: The Clarendon Press.

Rogers, R. (1992), 'London: a call for action', in Rogers, R. & Fisher, M. (1992), *A New London*, London: Penguin, pp xiii–xliv.

Roncek, D. & Maier, P. (1991), 'Bars, blocks and crimes revisited: linking the theory of routine activities to the empiricism of "hot spots"', *Criminology*, Vol 29, pp 725–853.

Rosenbaum, D.P. (1987), 'The theory and research behind neighbourhood watch: is it a sound fear of crime reduction strategy?', *Crime & Delinquency*, Vol 33 (1), pp 103–133.

Rosenbaum, D.P. (1988), 'Community crime prevention: A review and synthesis of the literature', *Justice Quarterly*, Vol 15, pp 323–295.

Rowe, C. & Koetter, K. (1975), 'Collage city', *Architectural Review*, August, pp 203–212.

Rowe, C. & Koetter, K. (1978), *Collage City*, Cambridge, Mass.: MIT Press.

Safdie, M. (1987), 'Collective significance', in Glazer, N. & Lilla, M. (editors), *The Public Face of Architecture*, London: Collier Macmillan, pp 142–154.

Safe Neighbourhoods Unit (1989), *Lighting up Brent*, London: Safe Neighbourhoods Unit.

Safe City Committee (1988), *Municipal Strategies for Preventing Public Violence Against Women*, Toronto: Safe City Committee.

Salusbury-Jones, G.T. (1975), *Street Life in Medieval England* (2nd edition), London: Harvester Press.

Schiller, R. (1988), 'Retail decentralisation: a property view', *The Geographical Journal*, Vol 154 (1), March, pp 17–19.

Schivelbrusch, W. (1988), *Disenchanted Night: The Industrialisation of Light in the Nineteenth Century*, Berkeley, CA.: University of California Press.

Schuck, M. (1996), 'Partnership needed to keep urban centres safe and secure', *Urban Environment Today*, July 11th, pp 12.

Schuster, J. (1995), 'Two urban festivals', *Planning Practice & Research*, Vol 10 (2), pp 173–187.

Schwartz, G.G. (1984), *Where is Mainstreet, USA?*, Westport, Conn.: ENO Foundation.

Scottish Office (1996), *CCTV in Scotland: A Framework for Action*, Edinburgh: Scottish Office.

Scottish Office Environment Department (1994), *PAN 46: Planning for Crime Prevention*, Scottish Office Environment Department/Scottish Office Central Research Unit, Edinburgh: Scottish Office.

Scraton, P. & Chadwick, K. (1991), 'The theoretical and political priorities of critical criminology', in Muncie, J., pp 284–298 op. cit.

Scruton, R. (1982), *A Dictionary of Political Thought*, London: Pan.

Scruton, R. (1984), 'Public space and the classical vernacular', in Glazer, N. & Lilla, M. (editors), *The Public Face of Architecture*, London: Collier Macmillan, pp 13–25.

Sennett, R. (1977), *The Fall of Public Man*, London: Faber & Faber.

Sennett, R. (1988), 'The *civitas* of seeing', *Places*, Vol 5 (4), pp 82–84.

Sennett, R. (1991), *The Conscience of the Eye: The Design and Social Life of Cities*, London: Faber & Faber.

Shaw, C.R. & McKay, H.D. (1921), *Delinquency Areas*, Chicago: University of Chicago Press.
Shaw, C.R. & McKay, H.D. (1931), *Social Factors in Juvenile Delinquency*, Washington D.C.: US Government Printing Office.
Shaw, C.R. & McKay, H.D. (1942), *Juvenile Delinquency and Urban Areas*, Chicago: University of Chicago Press.
Shaw, M. (editor) (1994), *Caring for Our Towns and Cities*, Nottingham: Boots The Chemists.
Sheffield City Council (1996), *Sheffield City Centre Business Plan*, March 1996, Sheffield: Sheffield City Council.
Sherman, L.W., Gartin, P.R. & Buerger, M.E. (1989), 'Hot spots of predatory crime: routine activities and the criminology of place', *Criminology*, Vol 27, pp 27–55.
Short, E. & Ditton, J. (1996), *Does Closed Circuit Television Prevent Crime? An Evaluation of the Use of CCTV Surveillance Cameras in Airdrie Town Centre*, The Scottish Office Central Research Unit, Edinburgh: Scottish Office.
Shotland, R.L. & Goodstein, L.I. (1984), 'The role of by-standers in crime control', *Journal of Social Issues*, Vol 40 (1), pp 9–26.
Simmons, M. (1996), 'Who's watching the watchers?', *The Guardian*, March 13th.
Simon, H.A. (1978), 'Rationality as process and product of thought', *American Economic Review*, Vol 8 (2), pp 1–11.
Skogan, W.G. (1990), *Disorder and Decline: Crime and the Spiral of Decay in American Neighbourhoods*, New York: The Free Press.
Smith, N. (1987), 'Of yuppies and housing: gentrification, social restructuring and the urban dream', *Environment & Planning D: Society & Space*, Vol 5, pp 151–172.
Smith, N. (1996), *The New Urban Frontier: Gentrification and the Revanchist City*, London: Routledge.
Smith, S.J. (1986), *Crime, Space and Society*, Cambridge: Cambridge University Press.
Social Research Associates (1995), *Public Perceptions of Nottingham City Centre*, Report for Development Department, City of Nottingham, Leicester: Social Research Associates.
Solesbury, W. (1993), 'Reframing urban policy', *Policy and Politics*, Vol 21 (1), pp 31–38.
Sorkin, M. (editor) (1992), *Variations on a Theme Park: The New American City and the End of Public Space*, New York: Hill & Wang.
Southworth, M. & Parthasarathy, B. (1996), 'The suburban public realm I: its emergence, growth and transformation in the American metropolis', *Journal of Urban Design*, Vol 1 (3), pp 247–263.
Stahura, J.M., Huff, C.R. & Smith, B.L. (1980), 'Crime in the suburbs: a structural model', *Urban Affairs Quarterly*, Vol 15, pp 291–316.
Standing Committee on Urban Planning, Housing and Public Works (1989), *Women and the City: Report of the Committee on the Problems of Women in an Urban Environment*, Montreal: Standing Committee on Urban Planning, Housing and Public Works.
Stanko, E.A. (1985), *Intimate Intrusions: Women's Experiences of Male Violence*, London: Routledge and Kegan Paul.
Stanko, E.A. (1990), 'When precaution is normal: a feminist critique of crime prevention', in Gelsthorpe, L. & Morris, A. (editors) (1990), *Feminist Perspectives in Criminology*, Buckingham: Open University Press.
Stansfield, K. (1995), 'Big brother or big father?', *Public Service and Local Government*, February, pp 7.
Stansbury, M. (1994), 'Getting to grips with the town centre', in Shaw, M. (editor) (1994), *Caring for our towns and cities*, Nottingham: Boots The Chemists.
Stickland, R. (1996), The twenty-four hour city concept: an appraisal of its adoption and its social and economic viability, unpublished MA dissertation, Department of Urban Planning, University of Nottingham.
Stollard, P. (1991), *Crime Prevention through Housing Design*, London: E & F N Spon.
Strathclyde Regional Council (1995), *Glasgow City Centre Public Realm Strategy and Guidelines*, Glasgow: Strathclyde Regional Council.

STRIDE (1995), *Safety Check Kit STRIDE Safe Travel For Women*, Nottingham: Women's Centre Nottingham.

Sudjic, D. (1996), 'Can we fix this hole at the heart of our cities?', *The Guardian*, Saturday January 13th, 1996, p 27.

Tein, J.M., O'Donnell, U.F., Barnett, A. & Mirchandani, P.B. (1979), *Street Lighting Projects: National Evaluation Programme Phase One Report*, Washington DC: National Institute of Law Enforcement and Criminal Justice, Law Enforcement Assistance Administration, US Department of Justice.

Tiesdell, S. & Oc, T. (1994), 'Re-creating the city', *The Surveyor*, 14th April, 1994, p 24–26.

Tiesdell, S., Oc, T. & Heath, T. (1996), *Revitalising Historic Urban Quarters*, London: Butterworth-Heineman.

Tilley, N. (1992), *Safer Cities and Community Safety Strategies*, Police Research Group, Crime Prevention Unit Series Paper 38, London: Home Office.

Tilley, N. (1993a), *Understanding Car Parks, Crime and CCTV: Evaluation Lessons from Safer Cities*, Police Research Group, Crime Prevention Unit Series, Paper No 42, London: Home Office Police Department.

Tilley, N. (1993b), 'Crime prevention and the safer cities story', *The Howard Journal of Penology & Crime Prevention*, Vol 32 (1), pp 40–57.

Toronto Transit Commission; Metro Action Committee on Public Violence Against Women and Children (METRAC), & Metropolitan Toronto Police Force (1989), *Moving Forward: Making Transit Safer for Women – A joint study of security on the rapid transit system relative to sexual assaults*, Toronto: Toronto Transit Commission; Metro Action Committee on Public Violence Against Women and Children (METRAC), & Metropolitan Toronto Police Force.

Toseland, R. (1982), 'Fear of crime: who is most vulnerable?', *Journal of Criminal Justice*, Vol 10, pp 199–209.

Townshend, T. (1996), *Lighting, Crime and Safety: A Preliminary Study of Newcastle City Centre*, Working Paper, Department of Town and Country Planning, Newcastle University.

Trancik, R. (1986), *Finding Lost Space: Theories of Urban Design*, New York: Van Nostrand Reinhold.

Trasler, G. (1986), 'Situational crime control and rational choice: a critique', in Heal, K. & Laycock, G. (1986), pp 17–24 op. cit.

Trembley, P. (1986), 'Designing crime', *British Journal of Criminology*, Vol 26, pp 234–253.

Tredre, R. (1995), 'Pubs see light in battle for women', *The Observer*, 6th August, p 11.

Trench, S. (1991), 'Reclaiming the night', *Town & Country Planning*, Vol 60 (8), pp 235–237.

Trench, S. & Lister, A. (1990), *A Survey of Taxi and Private Hire Car Use in Nottinghamshire*, Research Paper, Institute of Planning Studies, University of Nottingham, October 1990.

Trench, S., Oc, T. & Tiesdell, S. (1992), 'Safer cities for women – perceived risks and planning measures', *Town Planning Review*, Vol 63 (3), pp 279–296.

Trickett, A., Osborn, D., Seymour, J. & Pease, K. (1992), 'What is different about high crime areas?', *The British Journal of Criminology*, Vol 32, pp 81–89.

Urban Land Institute (ULI) (1983), *Revitalising Downtown Retailing – Trends and Opportunities*, Washington D.C.: ULI.

Urban Environment Today (1996a), 'Nine partners join Coventry on city management board', *Urban Environment Today*, July 11th, p 1.

Urban Environment Today (1996b), 'Ensuring city centres really mean business', *Urban Environment Today*, July 11th, pp 8–9.

Urban Villages Group/Aldous, T. (1992), *Urban Villages: A Concept for Creating Mixed-use Urban Developments on a Sustainable Scale*, London: Urban Villages Group.

URBED (1996), *Town Centre Partnerships: Their Role, Organisation and Resourcing* (Interim Summary Report), April 1996, London: URBED/ATCM.

URBED in association with Comedia, Hillier Parker, Bartlett School of Planning University College London, and Environmental and Transport Planning (1994), *Vital and Viable Town Centres: Meeting the Challenge*, London: HMSO.

Valentine, G. (1990), 'Women's fear and the design of public space', *Built Environment*, Vol 16 (4), pp 288–303.

Vidal, J. (1994), 'Darkness on the edge of town', *The Guardian 2*, 17th March, pp 2–3.

Wacjman, J. (1991), *Feminism Confronts Technology*, Oxford: Polity Press/Basil Blackwell.

Walker, N. (1991), *Why Punish?*, Oxford: Oxford University Press.

Walklate, S. (1996), 'Community and crime prevention', in McLaughlin, E. & Muncie, J. (1996), pp 293–331 op. cit.

Walsh, D.P. (1981), Book review of *Designing Out Crime* by Clarke, R.V.G. & Mayhew, P., *British Journal of Criminology*, Vol 21 (2), pp 189–190.

Warr, M. (1985), 'Fear of rape among urban women', *Social Problems*, Vol 32, pp 238–50.

Warr, M. (1990), 'Dangerous situations: social context and fear of vitcimisation', *Social Forces*, Vol 68, pp 891–907.

Warr, M. & Stafford, M.C. (1983). 'Fear of vitcimisation: a look at the proximate causes', *Social Forces*, Vol 61, pp 1033–1043.

Webb, B., Brown, B. & Bennett, K. (1992), *Preventing Car Crime in Car Parks*, Police Research Group, Crime Prevention Unit Series, Paper 34, London: Home Office.

Webb, B. & Laycock, G. (1992), *Tackling Car Crime*, Crime Prevention Unit, Paper 32, London: Home, Office.

Wekerle, G.R. & Whitzman, C. (1995), *Safe Cities: Guidelines for Planning, Design and Management*, New York: Van Nostrand Reinhold.

White, W.H. (1980), *The Social Life of Small Urban Spaces*, Washington D.C.: The Conservation Foundation.

Whitzman, C. (1992), 'Taking back planning: promoting women's safety in public places – the Toronto experience', *The Journal of Architectural & Planning Research*, Vol 9 (2), pp 169–179.

Wilkinson, P. (1993), 'City centres SWOT up on local needs analysis', *Planning*, 6th August, pp 14–15.

Williams, P. & Dickinson, J. (1993), 'Fear of crime: read all about it? The relationship between newspaper crime reporting and fear of crime', *British Journal of Criminology*, Vol 33 (1), pp 33–56.

Wilson, E. (1991), *The Sphinx in the City: The Control and Disorder of Women*, London: Virago.

Wilson, J.Q. & Kelling, G.L. (1982), 'Broken windows', *The Atlantic Monthly*, March, pp 29–38.

Wilson, J.Q. (1975), *Thinking About Crime*, New York: Basic Books.

Worpole, K. (1992), *Towns for People: Transforming Urban Life*, Buckingham: Open University Press.

Young, J. (1988), 'Radical criminology in Britain: the emergence of a competing paradigm', *British Journal of Criminology*, Vol 28, pp 289–313.

Zimring, F. & Hawkins, G. (1973), *Deterrence: The Legal Threat in Crime Control*, Chicago: University of Chicago Press.

Zimring, P. & Zuchl, J. (1986), 'Victim injury and death in urban robbery – a Chicago study', *Journal of Legal Studies*, Vol. 16.

Zukin, S. (1989) (Paperback edition), *Loft Living: Culture and Capital in Urban Change*, New Brunswick, New Jersey: Rutgers University Press.

Zukin, S. (1991), *Landscapes of Power: From Detroit to Disneyland*, Berkeley: University of California Press.

Zukin, S. (1995), *The Cultures of Cities*, Oxford: Blackwell Publishers.

Index